Fire Fighting

Fire Fighting

How It's Done

And How YOU May Have To Do It!

Robert L. Shrader

Library of Congress Catalog Card No. 96-61253

ISBN 1-56550-056-3

Printed in the United States of America

Cover graphics and design: Robert Brekke
Cover photograph courtesy of Dr. George McClary

Vision Books International
775 East Blithedale Avenue, Suite 342
Mill Valley, CA 94941
(415) 383-0962

Dedication

This book is dedicated to the millions of men and women of the past, present, and future who have served or will be serving their town, city, county, state, or country in one of the most hazardous of jobs—fighting fires—and particularly to those who work as either low-paid or completely unpaid volunteers.

Fire fighters are rarely given adequate thanks for the risking of their lives and health to benefit the citizens of their communities. We are fortunate indeed that some of our young people are willing to shoulder the responsibility of learning how to fight fires, and then to actively participate in fighting them, as well as to aid daily in many other kinds of emergencies. Volunteers who give so much of themselves in a spirit of unselfish devotion to the needs of others are truly our nation's most benevolent and generous unsung heroes. To these people this book is dedicated.

Disclaimer

Fighting fires is dangerous. Firefighters are professionals who are taught by other professionals how to deal with fires and with the many other emergencies to which they are called. Their training is ongoing and comprehensive.

The publisher has made this book available to help people become aware of what they might possibly do before the fire department arrives that might help save lives and buildings. The publisher and author assume no responsibility for any injuries or losses due to any of the introductory level statements made in this book about fighting fires and aiding in any other emergencies.

TABLE OF CONTENTS

PREFACE

This is a book for those whose blood may have been be stirred by the sound of the screaming sirens and yelpers, by the booming horns and clanging bells of fire engines—for the Walter Mittys among us who may have dreamed of putting out a roaring inferno and saving countless lives—and for those who may have wondered how such things are really done. It is for those of us who have wondered what is inside one of those shiny red, lime-green, or white fire engines that speed along our streets and highways. It is for a beginning or advanced "fire buff," he or she who has spent hours talking to firefighters, who has picked up quite a lot about what goes on at fire scenes, and who might like to learn more. Above all, it is for anyone who will probably be faced with a fire emergency at some future time—and that means EVERYONE!

You will learn how fire fighting has developed, what equipment was used and how it has been improved, the various ways by which fires can be attacked, and with what. You will find interesting, basic information everyone should know before attempting to put out a fire or aiding in other emergencies discussed in later chapters.

Much of the basic information given in this book will be of interest and possibly be very useful for all persons, particularly those living outside of well protected metropolitan areas. The information is also a once-over-quickly outline for anyone thinking of embarking on a career in the fire service. Every chapter could be expanded into a full book.

The basic information given here should also be known by: people working with civil defense organizations, field workers in utility companies, peace officers, security guards, state or national parks workers, forest workers, National Guard personnel, prisoners assigned to help fight forest fires, amateur radio operators providing emergency communications at wildland fires, Boy, Girl and Explorer Scouts, teachers and personnel of all levels of schooling, all home owners or renters, particularly if they have children—in fact just about everyone over eight years of age!

Fire protection methods in North America have evolved over a period of some three hundred years—beginning with volunteers and their hand-thrown buckets of water, upward and onward to hand-operated water pumpers. Today we have our modern, highly trained professionals using sophisticated fire fighting weapons of many types.

Techniques of fighting fires, from simple grass fires to highly involved forest fires, then smaller structure fires, to complex multi-story operations in larger cities are presented in simple terms.

The equipment of a small rural fire department may not be as impressive or complete as some found in larger city fire stations, but the background of fire fighting methods and the basic equipment used is fairly common to fire departments of all levels.

The approaches taken on fire theory, equipment, and fire suppression methods do not often delve very deeply into the more technical details of fire suppression, but even some experienced firefighters may find a few new views of some fire fighting fields.

The author has been involved for two decades as a fireman, board member, captain, training officer, and then chief of the rural Freestone California Fire Company. He became a member, secretary, and president of the board of the Twin Hills Fire Department. He was a member of "Volunteers in Prevention" with the California Department of Forestry for several years. Prior to his fire experience, he was a shipboard radio officer, a deputy sheriff, taught Merchant Marine Cadets at Kings Point Merchant Marine Academy for three years, and was a teacher for twenty-three years at Laney College in Oakland, California, where he started his writing career.

As is true with all nonfiction work, many people have contributed in one way or another to the final result. It is impossible to thank them all. Some of those to whom the author feels particularly indebted are Fire Chiefs Russ Shurah and James Voight, Fire Captains Roy Moungovan, Darrel Meade, Carl Magann, Ken McHugh, and George Drennon, and also to Dr. Jim Gex, Betty Boyce, Robert Dalzell of the N.Y. American Museum of Firefighting, and especially, his ever helpful and understanding wife, Dorothy.

Robert L. Shrader
1997

INTRODUCTION

The occupation of firefighter is always a rewarding one. The everyday work of firefighters is usually involved in saving and protecting lives and property. In their everyday work, firefighters, besides answering fire calls, are responding to a variety of accidents, providing first aid before an ambulance arrives, and making accident scenes safe. If a gasoline truck overturns, it is imperative that any spilled substances be contained and made less dangerous. This is usually up to a fire department. How this is accomplished requires more than a little knowledge and training.

It is natural that everyone at a fire scene would like to help, but too often untrained aid can be more of a hindrance than a help. Everyone should know something about how fires can be put out because we will all probably have a fire in a home or work area sometime. Will you know what to do?

A primary factor in fires and most emergency situations is time. In a few seconds, a fire can spread to the extent that only a well-trained and well-outfitted fire fighting team will be able to cope with it. If you see a fire start, how should you react?

Fire departments would like everyone to understand the importance of notifying them as soon as possible. If you smell smoke but cannot immediately find the source, telephone 911 or your area's emergency fire number right away! If you find the fire source and are able to put the fire out, you can always call back and the fire engines will be recalled by radio, although one of them will probably continue in to make sure that everything is actually OK. This is not a false alarm. Any fire department would rather be turned back after a few blocks of travel than to be called too late to be most effective.

In the case of a suspected or actual fire, call 911. Then make sure that everyone is out of the building or area. After that, if there is a fire, you may want to try to put it out with a fire extinguisher, or with water from a garden hose, or with a wet sack, or smother it with a tarpaulin, blanket, sand, or dirt. Which of these should you use for this particular fire?

In almost all cases, fires could be prevented. To aid in prevention of fires, most fire departments try to remind everyone in their districts of the danger of having loose papers around, of oily rags not in metallic containers, of overburdened electrical systems, of improperly fitted and installed stoves or fireplaces, and of dry grass and brush close around homes. Also, every home should always have an adequate hose and nozzle connected to a water faucet for use in an emergency. Is your home ready for a fire emergency?

Attics, basements, and garages are handy places to store great quantities of materials that can increase the possibility of a fire starting, or of increasing the severity of it. How about such places in your home?

In too many cases, homeowners add stoves, fireplaces, or electrical wiring which would never pass building code inspections. It should be understood that building codes are written specifically to reduce the possibilities of fires. Have you done any of these things to your home?

The duties of a fire department when fighting a fire, in order of importance, are to:

1. Protect lives
2. Give required first aid in emergencies of all types
3. Protect property
4. Put out fires
5. In most cases, help in the clean-up and securing of the property after a fire has been extinguished.

At a fire scene, the saving of lives is of primary importance. Surprisingly, though, it may be more important to take steps to protect adjacent buildings, called exposures, before the fire itself is attacked. After all, the exposures were not the things that were guilty of causing the emergency! In many cases, exposures may be deemed not to be in imminent danger, and the fire will be attacked immediately. Who makes this decision?

The concept of first protecting exposures is highly important in rural areas where grass, brush, or forest fires may be involved. All steps must be taken to protect homes and structures from an approaching fire. A decision must be made as to what action must be taken first. Who can make this decision? Past training, or more than a little fire knowledge, is important. This book briefly outlines what many people should know about fire and fire fighting, particularly if they are harboring the thought of possibly becoming a firefighter or volunteer firefighter some day.

1 Early Fire Fighting Equipment

The Begining Days

Captain John Smith, back in 1608, chronicled how his newly arrived Plymouth, Massachusetts, settlers tried to fight a fire that wiped out almost all of their settlement. This was probably the first written account of fire fighting in the New World. The early colonial fires were usually caused by either leaky chimneys or sparks flying onto the thatched roofs of homes and buildings. Since chimneys were often framed with wood and covered with clay and stones, they themselves could ignite when a hot fire in the fireplace over-heated them.

In order to provide some protection against fires, a law was passed that every colonial home must have:

1. *A ladder long enough to reach the roof*
2. *At least one leather fire bucket*
3. *A fabric swab or beater on the end of a 10- to 12-foot pole (somewhat similar to a floor mop with a long handle)*

Such beaters, when wet, could be very effective on small thatched-roof fires. If iron *grappling hooks* with a rope or chain attached to the unsharpened end were thrown over their small buildings, the whole thatched roof could be pulled off, or the entire building could be pulled down to provide a gap, or *fire break*, for an advancing fire.

To make water available for fire emergencies, each village was required to have a central water source. It might be a well, wooden tank, cistern, or a reservoir from which members of a *bucket brigade* handed filled water buckets from one person to the next along a continuous human line to the fire. The man at the end of the bucket brigade and closest to the fire tossed the contents of the buckets directly on the flames. If the fire had gained much headway before the brigade was set up, it took a brave and strong man to approach close enough to reach any of the flames with the water. The emptied buckets were then handed, person to person, back along a second line of people to the water source. If the bucket brigade was well trained, each person could pass along a full bucket with one hand while passing an empty bucket in the opposite direction with the other hand.

One effort to improve the effectiveness of such a primitive form of fire fighting was the development of a *hand syringe*. About a half-gallon of water could be siphoned up into its barrel. Two men held the barrel while a third pushed the plunger. They would be rewarded by the ejection of a thin stream of water for perhaps fifteen to twenty feet out of the nozzle

Another device, a portable *squirt*, operated on the screw theory. Water was loaded into a tank by buckets while one or more men at the back of the device turned the handle connected to a large screw (somewhat on the principle of a modern meat grinder). By constricting the nozzle outlet to a diameter of about 1/4 inch, some degree of internal pressure was built up, resulting in a thin stream of water forced out of its nozzle.

In 1666, an immense fire destroyed over 13,000 buildings over a five-day period in London, England. This acted as a spur to the development of much more effective fire-fighting and water-throwing equipment worldwide. The pumping machines that resulted were called *water engines*. One type of early-day fore-and-aft hand pumping engine, called a *hand tub*, is shown in Fig. 1-1.

A pumping engine of this type was pulled to the scene of a fire by two or more running men. A bucket brigade was set up to dump water into the wooden tank of the engine. As soon as water arrived, several men

on each pump handle alternately pulled down and pushed up on the pump handles, called *brakes*, pumping the water out of the tank and onto the fire.

The water discharge in the early water engines was fed up to and through a nozzle, called a *monitor* or *pipe*. The first nozzles were fixed and non-adjustable. They were designed only to throw water directly behind the engine and slightly upward. Later, the nozzle was made to swivel by using two rotary joints at right angles to each other on the pipe feeding the nozzle to allow the nozzle to be aimed at different parts of the fire. It was necessary to position the engine near the fire, fill it with water, aim the monitor, and then start pumping water.

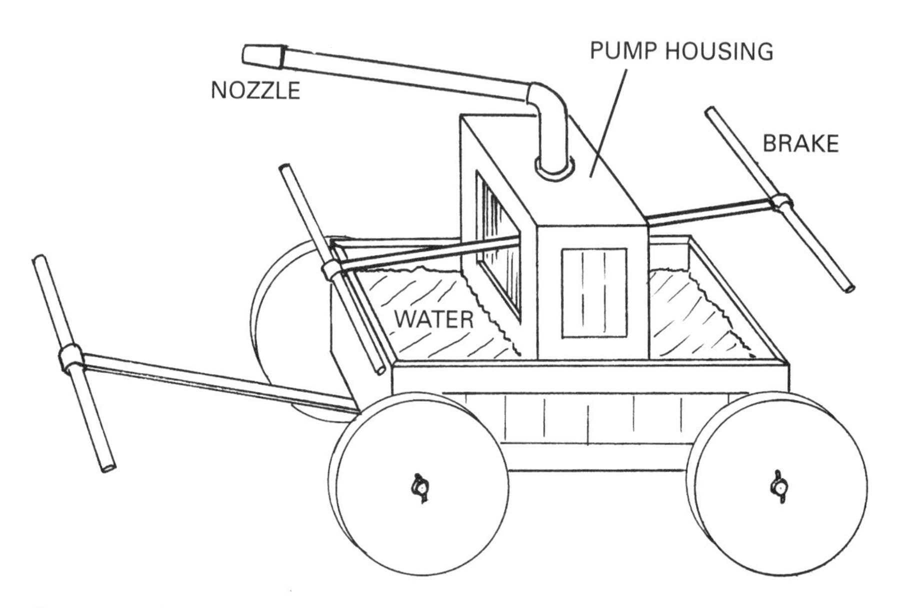

Figure 1-1. Form of early water engine or hand tub, pulled by men, filled by bucket brigade.

Basic Water Pumpers

Figure 1-2 represents the fundamentals of a somewhat more advanced side-pumping water engine, showing the left-hand brake coming down when the right-hand brake moves up. As the right-hand pump piston is drawn up, valve A opens, clapper valve B closes due to greater pressure on its left-hand side. The result is the drawing of water up into the right-hand pump cylinder. At the same time, the water-filled left-hand cylinder piston moves down, forcing clapper valve D closed, pushing valve C open, producing a discharge of water through the one-inch outlet pipe below. When the brake is pulled downward, a similar pumping action occurs but with the opposite cylinders, pistons, and valves.

To prevent the water from coming out in spurts as the cylinders were alternately discharged into the outlet, a metal air chamber or *dome* was included in the system, as shown. Any time a cylinder was discharged, it compressed some of the air in the dome. The compressed air could then force water into the outlet pipe when the pistons stopped momentarily at top and bottom. The result was a relatively steady output stream of water. Improved hand engines with ten men on each brake could throw a usable water stream 100 feet or more.

The first fire engine in the United States was imported from England in 1679 and was operated by paid firemen in Boston. It wasn't until about 1718 that the first volunteer *Mutual Fire Society* was formed to assist the town firemen. These first fire society men were mainly interested in providing protection for the homes and property of their own society members, or of the citizens who paid for insurance agreements with them. These volunteer fire societies, or *Benevolent Associations*, became very popular and wielded a great deal of political power. Many of the early paid fire departments gave way to all-volunteer fire companies. These companies were so powerful that their *exempt firemen* (five years active service) were often excused from serving on juries or in the militia. Many of these associations provided burial funds and financial aid to firemen's widows and children.

The first American-built fire pumper was constructed in 1733. It was horse-drawn and had its pumping brakes fore and aft (an *end-stroke* pumper) instead of on the sides (*side-*

stroke pumper). This particular machine never became popular because at about the same time, an Englishman, Richard Newsham, developed a very effective water engine. It was similar to previous models except that it was fitted with a stiff-walled, non-collapsible, sewn leather *suction* or *drafting* hose through which water could be sucked up into its piston pumps from any nearby body of water. This did away with the need for a bucket brigade if a water source was close enough to the fire scene.

By 1820, the monitor nozzles were no longer fixed on the pumping water tank. Brass outlet fittings were used, to which one or two 1-inch diameter, copper-riveted leather *leader* or *lead* hoses could be coupled. These allowed the firemen to lead, or transport, water to one or two separate areas of a building on fire. The result was a considerable advancement in water mobility and effective fire fighting.

A further development in engines was to feed water to the bottom of the two pump pistons' chambers through two pipes and their valves. These were fed by a 3-inch (or larger) leather drafting hose, either sewn or riveted around a stretched out 3-inch diameter steel spring. Such a hose would not collapse when it was used to suck up water from a pool. Water was drafted into the pump from a nearby water source, raised in pressure, and was then ejected through one or two smaller leader hoses. A single 2½-inch hose on such engines could carry about the same volume of water as six of the 1-inch hoses. These machines could either accept water from a drafting hose dropped into a nearby water source, or be coupled to street water hydrants, which were being installed in many cities in the early 1800s.

When water engines were pumped by hand, they rocked back and forth considerably. To reduce the rocking, many of them had side tanks filled with water to act as ballast. Other engines were fitted with external, hinged side support rods that reached to the ground to steady the engines while they were being pumped. Hand pumpers of this type were used up into the twentieth century.

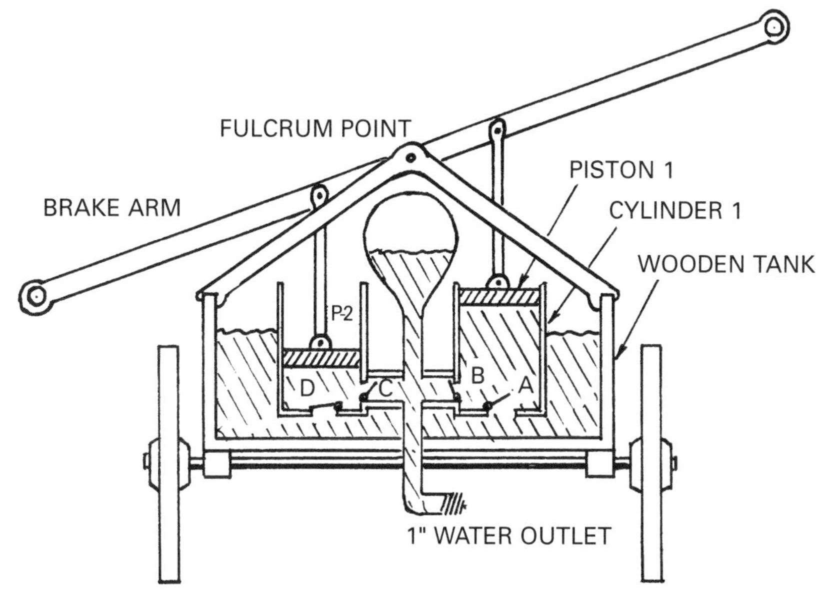

Figure 1-2. Early water engine pumps bucket-brigade water from its tank.

Side-stroke pumpers had their water intake at the rear and their hose outlet(s) in the front. End-stroke engines had their water intake fittings on one side and outlet(s) on the other side. Because the end-stroke front-to-rear wheelbases were considerably longer than the right-to-left wheel separation distance, the end-stroke engines were much steadier while being pumped. Their brakes could only be as long as the width of the carriage, limiting the number of men who could operate them. (The brakes on side-stroke engines could be the full carriage frame length.) To increase the number of men pumping end-stroke machines, some models were made with brakes at two levels, both fore and aft. One group of men pumped standing on the ground, and a second group pumped standing on fold-out platforms above the men pumping at ground level.

Figure 1-3 illustrates one way in which side-stroke engines could be constructed. The drafting inlet at the left is coupled to the pump. The output of the pump feeds past the pressure dome to the hose outlet(s). The brakes

were folded down alongside the engine when it was in motion. They were locked in a folded-out position while pumping, as shown. A pulling yoke, or two or more ropes, were connected to one end of the engine to allow it to be hauled to and from fires.

Dozens of manufacturers were involved in producing engines during the 1800s. Some of these hand pumping engines were still in use in rural areas up to World War I. Some may still be seen in parades, or in action at a water pumping contest, called a *Firemen's Muster*. They can throw water from a 2½-inch hose with a ¾-inch aperture nozzle 150 feet when pumped at top speed. At a reasonable pumping speed, ten to fifteen minutes of pumping is about all that any squad of men could be expected to work. After that, another squad would have to replace them. Engines of this type could easily pour a stream of water into a second or third floor window. Some of the larger engines, using 2½-inch hoses, could throw water 150 feet on the level. This could easily put usable streams into at least the fourth floor of a building.

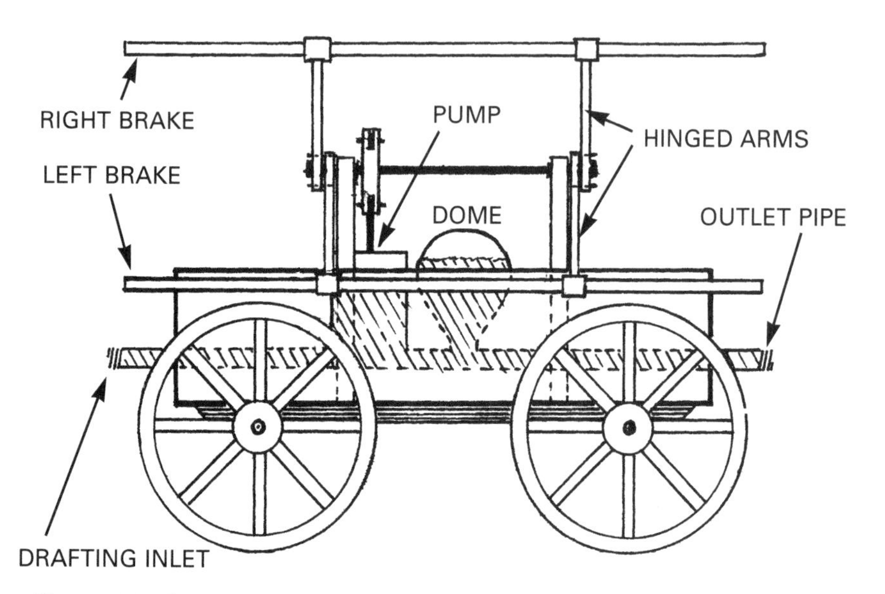

Figure 1-3. Layout of a pumping engine, drafting inlet, pump, dome and hose outlet. Brakes in folded-out position.

By the year 1776, New York City boasted of having 8 engines, 2 ladder trucks, and 170 firemen. By 1800, most larger cities had crews of appointed firemen who might be aided by volunteers. The days of hit-or-miss untrained volunteer firemen in cities was beginning to come to a close. By this time, hydrants had been installed in Philadelphia and other cities. The first ladder trucks with ladders up to 40 feet long and hose carriages capable of carrying up to 600 feet of hose would follow the engines. As early as 1809, the first fire boats were put into use to fight ship and dock fires.

In cities, paid groups of volunteer firefighters were being used, with a few full-time paid officers. These socially oriented fire societies should not be confused with our modern well-trained rural volunteer fire-fighters or the highly trained firefighters in our cities and towns today.

Steam Pumper Engines

Along with the development of steam railroad engines, a *steam pumper* fire engine was produced in 1830. Although heavy and slow moving, it used a steam engine to pump water. The volunteer societies saw this new improved engine as a threat to them. They thought it might make their organizations unnecessary. Because of such petty politics, steam pumpers were not put into the fire service for ten years after their initial development.

The early steam pumpers weighed over two tons and at first were hauled by twenty to forty men because no quick and simple means of using horses had been devised.

By the 1880s, most larger cities were using steam pumpers, usually drawn by three horses abreast, with a driver in the front seat and a *stoker* standing on a small platform behind the boiler tank.

When an alarm bell was sounded in a steam pumper station, the well-trained horses were released from their stalls in the back of the fire house. They galloped, without any guidance, into place in front of their own engine. Overhead harnesses with hinged collars

were lowered by ropes down over the horses' necks. The collars were snapped tight by the driver. He then climbed into the driver's seat and, hauling down on a rope over his head, pulled up the front door of the fire house. The horses dashed out, pulling the engine behind them. The horses could be out of the fire house within thirty seconds of the alarm. Steam could be up within about four minutes. With steam up, the stoker or engineer could open a valve and activate a vertically operating steam cylinder (similar to a driving cylinder of a steam locomotive), setting a water pump and a heavy flywheel into rotation.

The essentials of a steam fire engine are shown in Fig. 1-4. Kindling was always kept in readiness in the fire box below the boiler. The stoker riding on the back platform stuffed a bundle of kerosene-soaked excelsior (fine wood shavings) under the kindling wood and lit it. The excelsior and kindling rapidly ignited a special hot-burning *cannel* coal on top of the kindling. The heat from the fire developed steam from water fed to internal pipes in the boiler from a small water tank, often located beneath the driver's seat. The steam drove the vertical driving cylinder up and down, which in turn drove the water pump below it. Fire fighting water was led into the pumper system through a large drafting hose from a hydrant. It was then pumped out through the 2½-inch outlets shown below the engine. (Some engines had horizontal steam drives and pumps, but the theory of operation was essentially the same.) One end of the 20- to 30-foot stiff suction hose might also be held under the surface of a river or other water source to draft water into the engine.

The 2½-inch hoses, no longer leather, were now of woven linen or cotton, often with rubber linings. They were brought to the fire scene by a separate horse-drawn *hose cart* or wagon that followed the fire engine. A third vehicle, called a *hook and ladder truck*, usually from the same fire station, brought ladders, hooks on poles, and firemen. The hooks were originally used for pulling down or demolishing walls of small buildings. Today they are known as *pike poles*. They are 6- to 18-foot wooden poles with a metal hook and sharp point on one end, and are used something like a fireplace poker to push or pull burning boards in an active fire area or to pull down siding or ceiling sections.

As soon as the steam pumper arrived at the fire, the driver unhitched his horses and led them away until the excitement of the fire was over. Although the steam engines could not pump water much farther than some of the larger hand-operated engines, they were able to pump for hours at a time. A steam engine required a crew of only two or three men, whereas the large hand pumpers required 30- to 40-man teams, plus the firemen required to fight the fire.

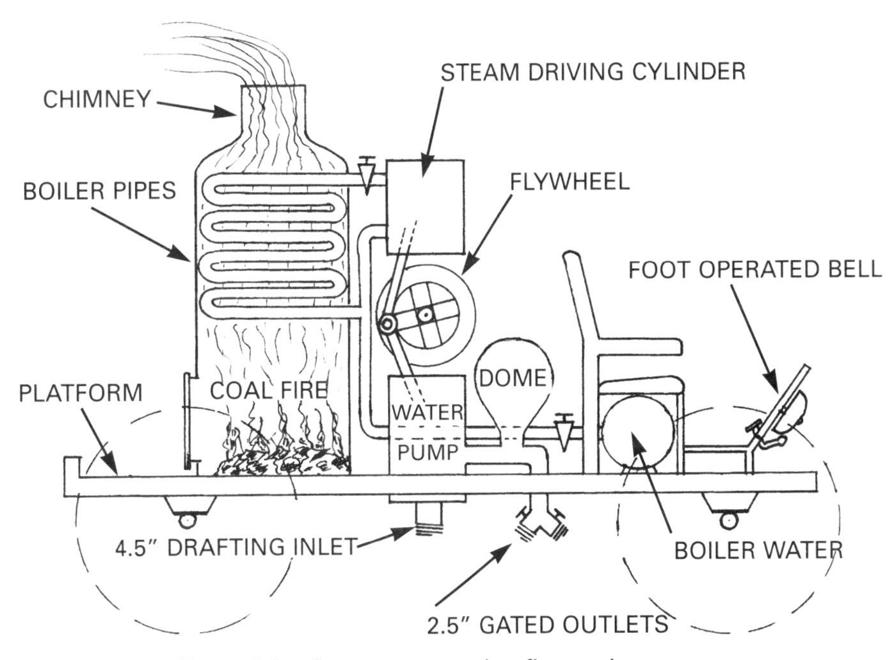

Figure 1-4. Essentials of a steam pumping fire engine.

Prior to the turn of the twentieth century, steam-engine propelled steam pumpers were used in a few cities, but were never popular. Whereas it took less than a minute for horses to move their fire engines out into the street, it took several minutes to get up enough steam to

start moving a steam-propelled pumper engine out of the fire house.

Steam pumper engines were manufactured up through the first decade of the twentieth century, with the last ones being used in the early 1940s. Many of these machines are still in working condition and could, in an emergency, be used for fire fighting today. You may find some of them in fire department museums in larger cities. Some may be restored and found on display in a few rural fire departments, and may show up at Firemen's Musters. Contact your local fire department to see if any musters are planned for your area in the near future.

Fire Equipment Development

By the 1870s, fire fighting had changed from the simpler volunteer departments with hand pumpers to more mechanized and modern operations. It was in 1873 that the Department of Agriculture first began to protect forests from fire. At about the same time, the *Hayes aerial ladder*, to the top of which a hose could be attached, was developed for metropolitan fire fighting. The *big stick*, as it was called, is still being used in one or more forms. The first models were raised and lowered by hand, but were later spring-raised. Some were operated by compressed air, and later by hydraulic pressure. The *fly*, or top section, of some of these aerial ladders could be extended to heights of 100 feet or more.

Hose elevators were also developed. These were upright posts with an elevated platform on which a fireman with a hose was stationed. It could be raised to heights of 55 feet. Crude wooden water towers could also be set up at the fire scene. They used two ladders tied together with ropes at the top to form an "A" assembly. With such a water tower, it was possible to direct water into higher floors of multi-story buildings.

In the 1880s, the first sprinkler systems were developed and installed in buildings. Alarm systems could be electrically connected to the local fire department headquarters. Such sprinklers were as much as 90 percent effective. They might leave only clean-up, or overhaul, work for the firemen when they arrived. One wonders why sprinkler systems were not made mandatory in all buildings many years ago. Today they are finally becoming required in larger structures and are being installed in many homes. Roof-top sprinklers are also being used on rural homes in particular, where brush and forest fires may blow burning embers hundreds of yards to a mile or more.

By the twentieth century, ladder trucks were so long that they required a *tillerman* stationed at the rear to turn the back wheels to guide the vehicle around corners. Life nets were used, although they are less successful than modern, relatively flat, air-filled bags that are used today.

Chemical engines were horse-drawn, water-throwing machines at first, but later were changed to gasoline-engine vehicles. They carried an airtight metal tank containing a water and sodium bicarbonate solution. A block or so before arriving at the fire scene, a fireman caused a container of sulfuric acid inside the tank to be poured into the bicarbonate solution. This generated carbon dioxide gas at high pressure in the tank. The pressure thus developed could force the solution through already charged (water-filled) one-inch hoses on hand-rotated reels to a distance of fifty feet or more. The watery solution came out of the hoses at several hundred pounds of pressure and was quite effective at suppressing small fires for a short period of time. If the proper proportions of alkali (sodium bicarbonate) and sulfuric acid were mixed, the solution should be neither alkaline nor acidic and theoretically harmless. Such speedy engines were often the "first in" to a fire scene and could get at least some water on the fire to start knocking it down. It was hoped that other fire equipment would arrive before the chemical engine's water solution was exhausted. Often the chemical engines were able to extinguish fires all by themselves if the fires were small.

By 1915, the trusty and greatly loved horses and the black-spotted Dalmatian dog mascots began to give way to the fast moving gasoline-engine trucks, termed *tractors*, that were used to haul the steam fire engines. Actually, the first all-gasoline operated fire engines were developed in 1910. Prior to this time, some gasoline-engine driven water pumps (to take the place of steam engines) had been manufactured but had to be pulled to the fire by men because the loud, unmuffled noise of those early gas engines frightened the horses. By the early 1920s, most fire engines and trucks were completely gasoline operated, opening what might be considered the modern fire-fighting era.

Centuries ago, in Dalmatia, a small country west of old Yugoslavia, black- or liver-spotted white "Dalmatian" dogs were bred to run alongside of coach horses to protect them from dogs, wolves, or other attackers. When fire pumpers became horse-drawn, it was only natural to use trained Dalmatians to protect the fire engine horses. Since the firemen had a lot of spare time, the dogs became their pampered pets. For many years after the last fire engine was pulled by horses, Dalmations remained the fire station mascot. They are still considered an emblem of firefighters.

2 Rural and City Fire Departments

Rural Fire Departments

To get some idea of what might be involved in a fire department that responds to your call, let's first take a look at a simple rural department. The make-up and outfitting of a rural fire department in an unincorporated area or in a small town depends on several factors. One is the size of the town or rural area, another is the number of businesses or farms from which volunteers can be recruited during the day, and a third is the wealth of the area. Other factors in the development of an effective fire protection unit are the concentration of population of a rural area, the citizen's concern for the area, and the ability of the fire chief and fire board to organize an effective fire fighting force.

The term *fire company* usually indicates a fire department that receives no operating funds from local donations. The term *fire district* may indicate a tax-based fire department. Whether a fire department operates from a fixed income-tax base, or whether it must rely mostly on donations from citizens of the area can make a great difference in its effectiveness and the type of equipment found in it. The various categories of fire fighting organizations today are typically: *city fire departments* backed by city taxes, *fire districts* backed by county taxes, *fire companies* operating mostly on donations, and *state or federal districts* operating on state and federal tax revenues.

In rural fire companies it is usually advantageous to pay one of the volunteers a retainer fee for daily part-time work at the station. Under the guidance of the chief, this volunteer assures that: Necessary paper work of the department will be kept current; schedules will be set up to see that volunteers are trained and are available when a fire alert is sounded; and that rolling equipment is ready to go at all times. Without a paid person, the fire chief and his officers might have to attend to these details themselves.

The Rural Fire Station

A possible architectural outline for a minimal rural volunteer fire station with no overnight facilities is shown in Fig. 2-1. At the right is an apparatus room that should be suitable for housing at least three fire vehicles, usually an engine, a water tender (previously known as a tanker, although this term is now used for fire suppression airplanes), and a pickup outfitted to operate as a brush truck. At the rear of the garage area should be space for work benches with locked storage shelving and cabi-

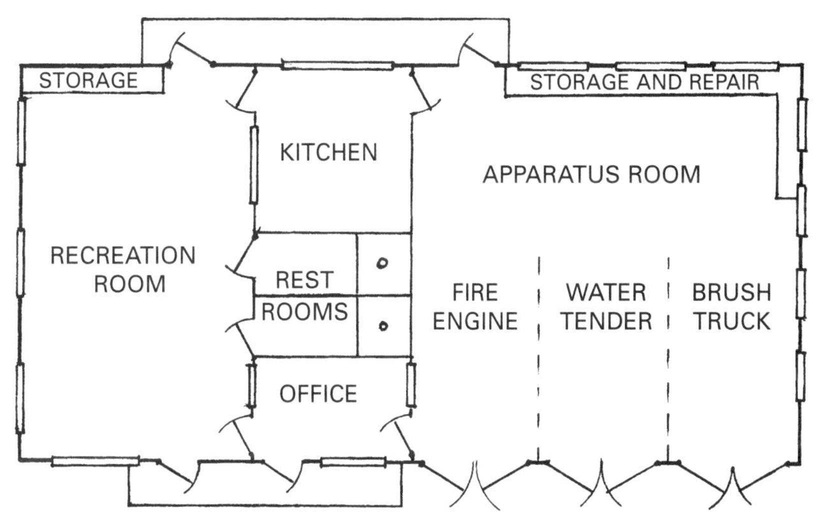

Figure 2-1. Possible requirements for a rural fire station.

nets. If a high-volume water hydrant is not available, there will have to be some kind of elevated water tank of 5,000 to 10,000 gallons capacity nearby to allow rapid gravity filling of the water tender, fire engine, and brush truck water tanks.

Rural fire companies are almost always in need of more money. They have had to devise ways to obtain it.

The inside end walls of the apparatus room may be used to hang up wet hoses to drain and dry, although a slanted ramp may be used instead. The higher the peak of the roof, the better for drying hoses. An outdoor rack may also be useful during dry weather. However, most modern hoses are plastic based and may be packed away wet because they will not rot from mildew.

The office space should contain a desk, chair, filing cabinet(s), telephone, and a fire-frequency radio receiver and transmitter. Good visibility in all directions should be provided for anyone working in the office.

Rest rooms are necessary in any firehouse, and showers are desirable for when the volunteers return from fires. If a department employs paid members and maintains 24-hour watches (unusual for small rural departments) it will be necessary to provide overnight facilities, perhaps on a second floor.

Because rural departments are often involved in community functions, a rentable activity area may pay for itself. For such activities, a kitchen large enough to allow several persons to work in it at the same time is important. The fire station garage may be temporarily cleared and rented out as a recreation or meeting hall, with the proceeds going for the fire department's operation. In some cases, it might pay to build a special activity room to rent out for community functions.

Water Supply

Water supply is of extreme importance to any fire department. Rural departments may have to depend on only the water that they haul to a fire in their own vehicles. There are usually no hydrants out in the country. Water for the working engines may have to be drafted from nearby streams, lakes, swimming pools, tanks, or wells.

Most rural home water systems today are driven by low-powered electrical pumps, producing up to perhaps 60 pounds per square inch (psi) of pressure with a storage tank of 50- to 100-gallons capacity. A single 1½-inch hose can drain such a system down to nothing in less than a minute. Systems of this type are useful for suppression of only the smallest of dwelling fires.

In many rural areas, all newly built structures may now be required to provide storage tanks with a capacity of 1,000 to 5,000 gallons of water on the property for fire fighting use (swimming pools may qualify if they are accessible to fire apparatus drafting hoses).

Vehicles

In some rural departments, more than three vehicles will be housed at the fire station. In some rural cases, as newer equipment is obtained, the older fire engines may not be kept at the main station but may be stationed at strategic locations in the surrounding district, perhaps at the homes of officers or firefighters. When alerted, volunteers manning such vehicles can receive their instructions regarding the location of the fire or other emergency by the two-way radio equipment in all modern vehicles. Volunteers, and particularly the officers, may also have issued to them small walkie-talkies or one-way alerting receivers (pagers) that they should carry with them at all times. Some member of the fire-fighting personnel may also have a handheld cellular telephone to call for aid from other agencies not available by the fire radio circuits.

Equipment

The backbone of any rural department's fire suppression ability is probably its 500- to 1,000-gallon, low-pressure (±200 psi) pumper or fire engine. Besides the main hoses, it may also carry one or two hose reels with 150 to 200 feet of heavy-walled rubberized fabric *hard-line*

hose, having either a ¾-inch or 1-inch inside diameter (ID). These hoses are normally kept charged (filled) and are fitted with combination straight-stream/spray/fog/off-type nozzles attached to them. Such hard-line hoses can be put into immediate operation when arriving at a fire. By the time one of the firefighters can run the hard-line hose to the fire, the driver should be able to get the engine's water pump started so that when the firefighter opens the nozzle valve, the water is ejected at once.

Besides the hard-line system, a pumper may have perhaps 20 lengths (normally 50 feet per length) of 1½-inch ID rubber-lined fabric or polyester covered hose, plus 10 lengths or more of 2½-inch ID *soft-line*. If, at the end of 500 feet of 2½-inch soft-line hose a *gated reducing wye* (on/off switchable Y) connector (with a 2½-inch inlet and two 1½-inch outlets) is used. The resulting lines are two 1½-inch hoses 1,000 feet from the pumper. For grass fires, another 500 feet to 1,000 feet of 1-inch lightweight soft-line may be carried. These are only example values. Each jurisdiction may have its own requirements as to the minimum number and lengths of hoses on fire engines, water tenders, and pick-up trucks. States and cities may also specify their own requirements.

Besides the water handling equipment, fire engines ("pumpers") should carry at least carbon dioxide, foam, and dry powder fire extinguishers for small or special requirement fires, a first aid kit, face masks with portable air tanks for each firefighter, plus two extension and one roof ladder. Structure and grass fire fighting tools, such as axes, shovels, and backpack extinguishers (discussed later) are some of the other accessories that may be carried on the rolling apparatus.

Two-way *vehicle-to-base* and *vehicle-to-vehicle* radio equipment is a necessity for all vehicles. Handheld transceiver (transmitter-receiver) radios for the chief and all officers at the fire scene are highly desirable.

An important rural fire department vehicle is a 500 to 2,000+ gallon *water tender*. It should carry at least 500 feet of 2½-inch, plus 250 feet of 1½-inch, ID soft-line. Its water pump should be capable of moving at least 200 gallons per minute (gpm) at the end of its 2½-inch hose line. A reel of 150 feet or more of ¾-inch or 1-inch hard (sometimes called *red*) line with a combination nozzle (solid stream/spray/fog/off) will also give the water tender some fire suppression capability if it happens to be first in at a fire. Water tenders should also carry two or more fire extinguishers, back-packs, a first aid kit, and perhaps 20 feet or more of large diameter (3- to 6-inch) hard-line for drafting water into its tank from lakes, swimming pools, tanks, or streams. Such a hard-line will not collapse when suction is put on it by the drafting pump, but it does not bend easily and is usually quite hard to handle. *Engine-tenders* are combination pumpers and water tenders, carrying up to perhaps 1,800 gallons of water, and are excellent all-around fire engines.

Heavy fire equipment presents some difficulties in rural fire fighting.

The primary function of a rural water tender is to refill the water tanks of fire engines at the scene of a fire, allowing an uninterrupted water attack to be maintained on the fire. When it is pumped dry, the water tender's waterline to the fire engine is uncoupled, taken to the closest water source to be refilled, then returned. A water tender, when filled, is very heavy and may never leave the road-way, particularly in wet weather. It may have to have its fire engines refilled through several hundred feet of 2½-inch or larger soft-line. If required to approach a fire that is down a steep country road, a water tender should probably be backed down to its fire engine so that it can be driven out forward when it goes for more water.

A third piece of apparatus for a rural department might be a 4-wheel drive, off-road radio-equipped pickup, carrying perhaps 50 to 200 gallons of water in a tank in its bed, along with a small water pump. It may also carry a reel of at least 150 feet of ¾-inch- or 1-inch charged hard-line. A vehicle of this type is very useful for spraying water as it runs alongside grass-fire lines. It should have a full complement of grass-fire tools discussed later, as well as several different types of fire extinguishers.

It may also carry a 120-volt ac generator, more accurately termed an *alternator*, used to power flood lights at night or to operate the motor of a *smoke ejector fan* when used at structure fire clean-ups. Because it is relatively light and maneuverable, it is usually first in at a rural fire scene and may carry the fire chief or another officer who is acting as the *incident commander* at a fire scene.

City Fire Departments

As the size of a town increases, so does the complexity of its fire protection and fire fighting system. The number of fire houses increases. In many smaller cities, the firefighters may be mostly volunteers, but they may be paid a small monthly wage for each training session they attend. More complicated equipment is required for high-rise buildings, as will be covered in later chapters. *Ladder-carrying* trucks become necessary to offer access to multi-story buildings, which are normally not found in many small rural districts.

Water Supply

Large quantities of water must be made available to engines from street water hydrants in bigger cities. City fire engines can pump their own tank water for the first minute or so. Hoses will be coupled to them from one or more fire hydrants within a short time after arrival at a fire scene. The low pressure (perhaps 60 psi) but high volumes of water from street hydrants through large hoses can be greatly increased by running the water through a fire engine's high-pressure pumps (developing at least 200 psi). This is particularly important when fighting very hot fires or fires several stories above street level (water pressure is lost at 0.43 psi per foot of elevation).

Special Trucks and Equipment

Some city departments may have special hose trucks that carry many hoses of a variety of sizes. Another special truck may carry only chemicals and mixing machinery to produce great quantities of a foam-type fire extinguishing substance that is very effective on many types of fires. Some trucks may be small, electrical generating plants on wheels, fitted with large floodlights plus electrical outlets to operate other equipment. Since pumpers and other apparatus are operated by gasoline or diesel engines, fuel trucks must be sent to extensive fires to refuel the equipment. Field kitchens may be required to feed the fire-fighters when a fire lasts for a long period of time. Department ambulances, resuscitators, and equipment used to refill depleted air-mask air tanks or operate air-pressure type tools may be required.

Personnel

Rural

The personnel of a rural volunteer fire company consists of a chief, one or more captains, one or more lieutenants, and the firefighters. In the past, all firefighters were men, but today, when properly trained, some women have become excellent firefighters. Backing up the firefighters and officers is usually an elected district fire board or commission made up of local citizens. Such a board usually includes at least one of the officers of the company, or the chief may possibly act as an *ex-officio* member. The fire board determines basic operating policy and is responsible for money spent by the department. The officers and firefighters may be either paid members, partially paid members who may be reimbursed for time spent in training,

or unpaid volunteers. It has been found poor procedure to pay firefighters for only those fires to which they respond, because some firefighters have been known to start a fire somewhere so they can be paid to help put it out!

When a rural company is on an active fire or emergency call, one of the firefighters, or perhaps a trained auxiliary member, should be at the fire station radio controls. This is important if additional fire fighting help must be called in, or if doctors, ambulance, police, etc., must be summoned by telephone.

An auxiliary can be most helpful for rural departments. Besides one of their members possibly operating the radio station, they can bring sandwiches and coffee to the firefighters at fires of long duration. They are very valuable helpers when putting on periodic breakfasts or benefits to provide money to run smaller rural departments.

City

Unlike rural departments, the fire chief of a large city department may not go to any fire unless it is a multiple alarm type. Larger cities may be laid out into several fire districts, each having its own *battalion chief* and a full complement of officers, firefighters, apparatus, and fire stations.

To train its personnel adequately and maintain an up-to-date fire fighting department, a larger city may have its own fire school. Training may also be held in conjunction with other nearby cities, or with county, state, or federal government services. Community and other colleges may give fire science courses. Members of rural departments may take training at fire schools in nearby cities. Cities lying adjacent to one another have learned that it pays to have firefighters of all nearby fire departments with similar fire training backgrounds. When a fire becomes a conflagration, as in the 1991 Oakland hills fire in California, a great many mutual aid firefighters and apparatus are required from dozens of surrounding departments. It is important that all fire personnel at a fire use similar operating procedures. All fire hose fittings and other equipment should be interchangeable for maximum fire extinguishing efficiency.

There are three things that should be understood by anyone who may ever fight a fire: What "fire" is, what may start it, and how it may be stopped. Let's take a look into the interesting subject of fire in our next chapter.

3 What is Fire ?

Requirements for Fire

Whether you like to think about it or not, someday you may be called upon to extinguish a fire. You should know something about what fire is, what methods can be used to stop it, and better yet, what we can do to prevent fires from starting. These are intriguing subjects. This chapter may seem a little technical in some parts, but only because heat and fire are complicated subjects. It has been simplified as much as possible, and covers only the basics.

Since the beginning of the science of fire fighting, it has been known that most common fires are due to the heat caused by a chemical reaction that occurs when wood is oxidized. *Oxidation* is the chemical combining of oxygen with any of a variety of substances that can act as fuels. The air we breathe is about 21 percent highly active oxygen gas, 77 percent chemically inert nitrogen gas, plus small amounts of water vapor and a few rare gases. Of course, merely allowing a fuel, such as wood or paper, and air-oxygen to come in contact with each other is not enough to start a fire. Something else must happen to get the chemical reaction started.

The normal, very slow oxidation, or combining of air-oxygen with wood, results in a graying of the wood's color. This occurs on the exposed surface of any board. It may require days or months for this effect to be noticed. If dry wood is heated in the presence of the oxygen in the air, at some temperature between 400° and 500° F, the wood will burst into a chemical reaction that produces wood gases and visible flames. The burning gases will immediately have a temperature somewhere in the range of 1,600° to 1,800° F. Heating any substance speeds up its chemical reaction. Chemical reactions produce heat, although some produce more than others. The heat of the rapid chemical oxidation that is responsible for flaming gases will radiate heat to any other nearby wood or fuel. This in turn develops oxidation-produced hot gases and flames in the nearby fuels, allowing the fire to spread rapidly. If the oxygen content of air were to be increased in some way, the burning rate would be increased greatly.

To produce fire in most materials, three things are required:

1. *Fuel*
2. *Oxygen*
3. *Heat*

In the next chapter it will be pointed out that if you want to extinguish a fire, (1) the fuel might be removed, or (2) oxygen might be prevented from coming in contact with the heated fuel, or (3) the base of the fire might be cooled. From this reasoning comes the traditional firefighter's fire triangle, illustrated in Fig. 3-1. To start a fire, fuel, oxygen, and heat are required. To put out a fire it is only necessary to interfere in some way with any one of the triangle's fuel-oxygen-heat sides. More about this later.

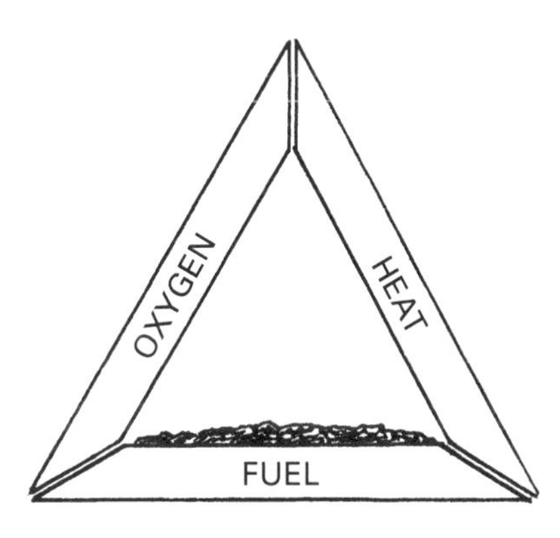

Figure 3-1. Fire triangle: All sides required for a fire. Remove any side to extinguish the fire.

How Fires Start

Energy can be transformed from one type to another under certain conditions. When you energetically rub your two hands together the energy of pressure and motion, or friction, converts to infrared or heat energy in the molecules of your skin. Your nerves sense or detect the heat. If two sticks are rubbed together, heat, or infrared radiation, is produced, heating those parts of the sticks closest to where the frictional energy is being applied. If a dry stick is rotated against one spot on a piece of dry wood, enough frictional energy may be radiated to the hot wood dust produced to combine with air-oxygen and develop a fire.

In such a way primitive people (and Boy Scouts) could start camp fires. Steel striking steel can cause hot particles *(sparks)* to fly from the point of contact to start fires.

Light rays and other waves from the sun can be focused to a small spot on a fuel with a magnifying glass. By concentrating infrared heat and light-wave energy on a very small area, molecules at that point can be brought up to their kindling (oxidizing) temperature. Heated fuel and air-oxygen chemically combine, and a fire starts. Glass bottles lying on dry grass in hot summer sunlight have been blamed for starting fires in this manner.

Energy carried by electrons, tiny electrical particles that can be pushed along a copper wire (forming an electric current) and through the tungsten-wire filament of a light globe will cause frictional heat as they progress through the resistance presented to them by the filament. The energy carried by the current might only be enough to heat the filament so that it

Energy Radiation Frequencies

Just what is meant by fire and flames? Fundamentally, both are the result of the *radiation*, or throwing outward, of heat energy. While the subjects of heat, fire, and flames are really quite complex, a few simple ideas can provide an adequate groundwork for an understanding of what flames are and what happens in a fire.

Let's deviate a little here. It is reasonably accurate to say that certain types of energy, such as heat from the sun, is radiated in wave form. Different kinds of energy waves have different *frequencies.* But what is meant by frequency? If electric currents are made to go back and forth in a wire, or *alternate,* at a rate of 100 times a second, this may be expressed as either 100, or 10 x 10, or 10^2 cycles per second (cps). 100 cps may also be termed 100 *hertz* (abbreviated Hz). Radio broadcast stations radiate waves of alternating electric and magnetic wave energy at frequencies in the range of 1,000,000 Hz, or a million hertz (a *megahertz,* or MHz). This may also be expressed as 10^6 Hz. Our FM broadcast stations, some of the fire service two-way FM radios, and many TV stations radiate electromagnetic energy waves in the 100,000,000 Hz (100 MHz, or 10^8 Hz) range.

The *powers of 10,* such as 10^6 above, (which means "1 times 10 to the 6th power") are a mathematical shorthand that can be used to express large numbers such as the frequencies of heat, light, and radio waves. As examples, you know that 1 x 10 = 10, and also 1 x 10^2 = 100. To simplify things, 1 x 10^6, more simply written as 10^6 (the "1 x" is just understood to be there), is equal to 1,000,000 (6 zeros after the number). If you write 2 x 10^9, you are specifying a very large number, 2,000,000,000, and you have saved writing quite a few zeros. Each additional power-of-10 numeral adds another zero to the number being described.

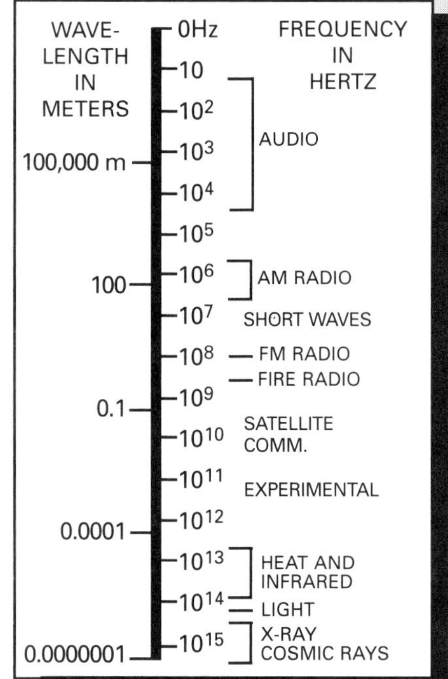

Figure 3-2. Electromagnetic frequency spectrum in frequency with some equivalent wavelengths shown.

Heat (basically *infrared* electromagnetic waves) represents energy having frequencies somewhere between 10^{13} and 10^{14} Hz. (That's really a lot of zeros after a 1.) Above this frequency comes a narrow band (near 10^{14} Hz) which we call *light frequencies,* because our eyes are sensitive to them only. Still higher frequencies are *ultraviolet* (too high a frequency to be sensed by our eyes), then *X rays* and then *gamma rays* (10^{15}+ Hz). The whole electromagnetic spectrum can be laid out in frequency as shown in Fig. 3-2.

The same energy spectrum in frequency can also be laid out in *wavelengths,* or how far an ac wave travels through space during the time it takes to go through one full cycle (zero to positive, to zero, to negative, and back to zero). The highest fre-

radiates in the 10^{13} Hz infrared heat range. If more current is made to flow, the filament will heat more and radiate a wider band of energy, including the light frequencies. The filament will have become hot enough to radiate both heat and light frequency energy which is adequate to start fires in many cases.

Bare wood will absorb any heat (infrared) energy directed at it. This is known as pyrolysis, the subjecting of heat to a material and the resulting change produced. If the wood is in a confined area where little or no air can circulate around it, the wood may continually absorb the radiant energy being applied to its surface. Should the captured heat energy increase the surface temperature and dry out the area enough to support a fuel and air-oxygen chemical reaction, a fire will be kindled at that spot. This can happen if a bare copper or iron hot-water pipe is allowed to lie or press against the edge of a hole in a wooden wall or building stud. Many structure fires have been started in

quencies will have the shortest wavelengths. The formulas to convert any frequency in hertz to its wavelength in meters, and vice versa, are:

wavelength = 300,000,000 ÷ frequency
frequency = 300,000,000 ÷ wavelength

The number "300,000,000" (3×10^8) is the *speed of light* (or any electromagnetic wave in space) in meters/seconds (m/s). Therefore, the wavelength of a 100 MHz FM broadcast radio signal is 300,000,000/100,000,000, or 3 meters. The radio wave travels 3 meters away from its transmitting antenna in the time it takes to complete one of its ac type cycles. Heat, light, and radio radiation may be discussed in either wavelength or in frequency. Both tell the same story but from different viewpoints. We will use frequency here, only because it will be used in discussions about fire radio communications later. (Most fire training uses wavelength.)

The sun radiates a wide band of energy waves. When infrared sunlight energy strikes the molecules (tiny particles) that make up your skin, this energy re-radiates from the surface molecules to those in the next lower layer, and so on. When the re-radiated energy strikes a nerve ending located a short distance below the skin surface the nerve acts as a receptor. It chemically/electrically signals your brain that heat frequency energy is being received at that part of your skin.

Somewhat similarly, light frequency energy striking the light receptors in your eyes will signal your brain that light frequency energy is present at the back surface of your eyeballs. Different parts of your eye receptors respond to different light frequencies, allowing you to see different *colors* (frequencies). We should be glad that our eyes are not sensitive to the lower infrared frequencies, or we would see a cup of hot coffee as a glowing object, somewhat as you see glowing red coals in a fireplace or barbecue pit

It might be mentioned that audible sound waves are not electromagnetic, but are waves developed in the air by alternate compression/rarefaction waves of the gases that make up the air. Your ears are receptors for this compression/rarefaction type of air wave provided the waves are at some frequency between about 20 and 20,000 Hz. Some young people may hear frequencies up to 24,000 Hz, but this ability is progressively lost with age. Sound is a slow traveling wave, moving at only about 1,100 ft/s (330 m/s) in air. It is interesting that fire can produce its own sounds. A candle flame radiates a sound wave that is close to the lowest threshold of human hearing (15 Hz). Heated wood, just before it bursts into flame, produces a very high-frequency molecular motion sound. It is possible that fire detection systems may be based on these facts in the future.

this manner. Even if the metal pipe temperature is no more than perhaps 150° F, if the radiant energy can not escape the area fast enough, the kindling temperature of very dry wood can be reached. Structural wood near hot brick chimneys with old, loose mortar often ignites for this reason.

In the fire service, burning fuels such as wood, wood products, cotton fabrics, and grass are known as Class A fires (usually extinguished best with water). The formation of rust on iron is a case of oxidation, but it occurs so slowly that its heat radiation is not detectable. When oxygen chemically combines with certain hot carbon-hydrogen-oxygen fibers such as those of wood (chemically $C_6H_{10}O_5$, where C is carbon, H is hydrogen, and O is oxygen), because of the localized heat energy, the chemical action liberates more energy than was used to start the chemically combining process. Nearby molecules are radiated with quantities of this increased heat frequency energy, which in turn pass heat energy to molecules adjacent to them. As a result, fire becomes a self-supporting, growing chain reaction.

Old brick chimneys must be checked continually to make sure they are fire safe.

If sufficient heat energy is applied to wood, air-oxygen begins a complicated chemical reaction. Some of the oxygen combines with hydrogen in the wood to form water (H_2O) vapor, and some combines with carbon in the wood to form carbon monoxide (CO) gas. The carbon monoxide, if enough air-oxygen and heat are available, may chemically combine with oxygen to form carbon dioxide (CO_2), producing still more heat energy with this second chemical-combining process.

Some of the results of burning, namely hot water vapor, hot carbon monoxide, hot carbon dioxide, and other hot molecules called *tars*, contain a considerable amount of energy as they float upward away from the seat of a fire as smoke. These molecules are hot enough to act as sources of heat radiation to other substances they pass. If their radiation continues long enough, nearby objects may burst into flame.

It is interesting that if stacked hay bales are a little damp, the chemical action of bacteria working inside the hay may produce warmed pockets which can accumulate enough heat to start the hay burning.

Another unusual formation of fire, called *Will o' the wisps*, are flames that sometimes shoot up out of a pool of water. These are the chemical result of phosphorus hydrides on the pool bottom combining with oxygen in the water to form a hot gas that rises to the surface and burns as a flame when it comes in contact with air-oxygen.

When a wood fire is allowed to burn to completion by feeding it all the oxygen it can use, the residue will be a white, powdery ash composed of combinations of the many trace chemical elements that were in the wood—silicon, sulfur, potassium, sodium, etc. The basic C, H, and O wood molecules have been either chemically combined with air, or have been given off as pure black carbon *soot* molecules, or have left in the form of gases and tars in the smoke.

If a wood fire is extinguished by cooling it with water, complete chemical reactions have not been allowed to occur. The residue is charred or blackened wood. The black is nearly pure carbon that has not been driven free nor had a chance to combine with oxygen and be liberated in the form of hot carbon monoxide, carbon dioxide, or other more complex gases.

The color of the smoke rising from a fire tells firefighters something about the fire beneath it. Dark smoke usually indicates involvement of petroleum-based or plastic materials, although wood also develops a certain percentage of brown tars. Light gray smoke suggests that the fire is basically dry wood. When the smoke turns white, it indicates the fire is being fought and is coming under control with steam or water vapor as a major part of the smoke. Pink showing below a smoke cloud indicates an active fire with a large flame base.

Conduction and Convection

We say that heat travels through solids by conduction, meaning by the re-radiation of

infrared energy from molecule to molecule from the surface down into any solid substance.

Heat can travel by *convection*, also. This means movement of hotter, more expanded and therefore lighter, gases upward through heavier cooler gases, such as air. The hotter a gas is, the more its molecules radiate energy waves, and the harder they push other nearby molecules away. If there are fewer molecules in a given volume of gas because they are pushing each other farther apart, the gas must be lighter. This is the reason why a hot air balloon rises—its hot air has its heated molecules farther apart—making its internal air-gas lighter than the cooler air-gas outside of the balloon. It is also the reason that the flame of a candle stretches upward from the wick. Its heated gases, hot enough to include some higher frequency orange light, are much lighter than the surrounding air, so the flame and any hot smoke from the wax moves upward.

The conduction of heat through a solid material is caused by energy radiation. The burning top surface of a wooden table first directly involves its top layers of wood molecules. These hot molecules radiate energy in all directions. The cooler molecules below expand when radiated from above. When the lower ones accept as much energy as they can, they begin to expand and radiate energy on their own, heating the molecules still further beneath them. Before long, the underneath surface of the table becomes warm to the touch. Therefore, we can say that the wood conducts heat through it. Since molecules must accept radiant energy and expand before they can re-radiate, the conduction of heat is far slower than radio, light, or heat radiation which travel at 300,000,000 m/s, or 186,000 mi/s.

As the wood molecules of a board accept more and more heat, they begin to break down into their constituent carbon, hydrogen, oxygen, and other atoms. The heated, expanding hydrogen gas drives itself and some carbon molecules up through the surface of the wood, forming a visible-frequency hot gaseous flame as they combine with air-oxygen. The combining of hydrogen gas and air-oxygen by heat energy forms water (H_2O) vapor.

Carbon monoxide is a highly active gas looking for something with which to combine. It may combine with blood platelets in the lungs, preventing them from delivering oxygen to the your body, which can result in your death. It is therefore a poisonous gas. Carbon dioxide is a stable, inert gas, and while it does not combine with blood platelets, it can smother a person by taking the place of oxygen in the lungs. For this reason those persons fighting a fire should not work inside burning buildings unless they are wearing a fresh-air supplying mask.

You have probably noticed that the conduction of heat through a wooden handle on a pot or pan is rather slow compared to the conduction of heat through a metal handle. Dry wood is a good heat wave *insulator*. This is caused by the air spaces that exist between the wood molecule fibers. Air and other gases have molecules that are much more widely separated than the molecules of solids and liquids. As a result, when gases are radiated by heat, they expand (as do metal molecules), but gas molecules have much more space around them, allowing them to expand greatly before internal pressure increases to the point that they begin to radiate energy from themselves.

The best heat-insulating materials are substances filled with many small air or inert gas pockets. There is no such thing as a perfect insulator that can stop the travel of heat. Eventually, heat travels through any wall, regardless of how thick it is or of what material it is composed.

Because a material is a good heat insulator does not mean that it will not burn. Dry wood is a good insulator but burns quite readily. Asbestos, another good insulator, will not oxidize and produce chemical reactions at normal flame temperatures of 1,600° to 1,800° F, so it does not burn.

The molecular structure of metals, solids, and liquids is such that there is relatively little space between molecules. The molecules can expand little when heated, and they therefore begin to re-radiate heat energy to adjacent molecules more quickly and easily than do gases. If sufficient heat energy is supplied, even metals can be made to melt or liquefy, and if still hot-

***Question:* What do you think would be the shape of a candle flame if it were inside a space vehicle that was in orbit around the earth?**

***Answer:* With no gravity, the candle flame could not float upward, since there is no gravitational up or down when anything is in orbit. The flame would expand outward in all directions, producing a round, ball-like flame, which would rapidly melt and consume the candle wax.**

Question: Why would a metal handle of a cooking pot at 120° burn your hand, whereas a wooden handle at 120° would not?

ter, their molecules will push apart so vigorously that they vaporize (become gases) just as water or other liquids do.

Some hot metals will combine with air-oxygen, while some will combine with other gases, such as chlorine. *Metal fires* are chemical reactions that may require no oxygen at all and cannot be stopped by flooding with water. In fact, water combines violently with cold metallic sodium, potassium, and lithium. They break down the water into hydrogen and oxygen atoms which, with the heat of chemical reaction, can burn fiercely, spewing out hot steam and metal particles. Application of reaction-stopping chemicals must be used to stop the heat chain reactions that are being produced in active-metal fires. These chemical reactions are too complex to be dealt with on the beginner level.

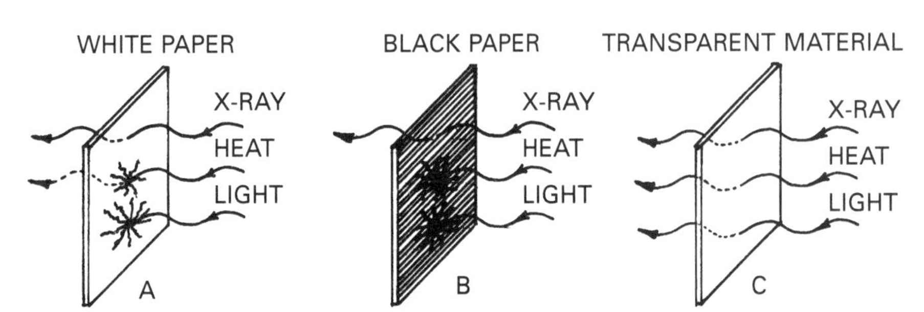

Figure 3-3.
(A) X-ray and light penetrate white paper.
(B) Only X-ray penetrates black.
(C) All penetrate most transparent materials.

As you continue to heat an object, the energy frequencies radiated from it may increase from the lower infrared energies to include the higher red, then orange frequencies, to the still higher yellow, and finally to the blue and violet frequencies. It might appear to become *white hot* when the total broad spectrum of all visible frequencies is radiated at the same time. Hot gases released from heated wood usually burst into flame, or active oxidation, before heating the wood to a red color. Red hot coals indicate the radiation of energy at a frequency higher than infrared, at low visible frequencies. As mentioned above, a tungsten metal filament in an inert-gas filled lamp bulb can go from infrared or black hot (900° F), to red hot (1,500° F), to orange hot (2,000° F), to yellow hot (2,300°) and finally white hot (2,800°) as its temperature is raised by increasing the electric current through the filament.

Transparent substances, such as glass, can accept radiant energy and immediately re-radiate it in the same direction with almost no attenuation. If the first surface of a sheet of glass is shiny smooth, it will not only transmit heat and visible frequencies through it but will reflect some of the energy at an angle equal to the angle at which the light is striking the glass surface. The sum of the transmitted plus the reflected plus any glass heating energy losses will be equal to the original, called the *incident energy* striking the glass.

Because transmission of heat and light energy through glass requires re-radiation of the energy from molecule to molecule, energy transmission through glass is slower than that of light frequencies in air, or in the vacuum of outer space.

It is interesting to note that *white paper* will pass X-ray frequencies through it with almost no loss, will transmit light frequencies through it to a limited extent, and will transmit lower infrared and heat frequencies poorly (Fig. 3-3a). It will intercept some light frequency energy and become slightly warm. With the same amount of a lower frequency infrared energy the paper may capture enough heat energy, to start burning if in the presence of oxygen.

Black paper will not attenuate or weaken X rays to any extent, but will stop both light and infrared frequencies (Fig. 3-3b). You probably know that a black surface warms much more than a white surface when exposed to light or heat. The molecules of a black paper will capture all of the radiant energy, which raises the temperature of the black surface. Incidentally, something is *black* only because it does not reflect or transmit any visible frequencies. Black surfaces capture more heat energy than white and are more likely to ignite.

Convection of heat is one of the problems in structure fires. Consider a fire to be in progress on the floor of a room. The smoke is hot gas—because its re-radiation of energy

Answer: Wider spacing of wood molecules prevents rapid loss of heat energy, so the skin is not bombarded with as much infrared energy in a given time.

pushes its molecules far apart, it is quite light. As a result, the hot smoke will be pushed upward by the heavier, lower, cooler air in the room. These hot convection currents of gas will mushroom outward along the ceiling from the point where they strike it. When the hot gases reach a wall surface, they tend to move down the wall, but are curled inward toward the center of the room by the lower, cooler, denser air. As a result, the air in the area near the ceiling is in a constant hot, turbulent motion. It is not long before such a ceiling surface can be brought up to kindling temperature.

Note that the coolest area in a room is always near the floor. If you are ever fighting a fire in a room and your ear lobes begin to tingle, drop to a crawling position immediately, and get out of there! In fact, anyone trying to get out of a room on fire should bend over and run, or drop to the floor and crawl to safety. Parents and school teachers should expound on this thought—**down and out!**

Anyone trying to get out of a room on fire should bend over and run, or drop to the floor and crawl to safety. Parents and school teachers should emphasize this thought— down and out!

Backdraft Explosions

If a fire starts in a room that is fairly well sealed, the normal 21 percent air-oxygen is soon reduced. When it is reduced to about 16 percent, there is insufficient oxygen for flames, and everything smolders. The room surfaces may become superheated, but with insufficient oxygen, they cannot chemically combine with anything and "burn." Eventually, with insufficient oxygen, the room will cool, and the fire will be out.

However, if a window is broken open or a door is opened while the room is in the hot smoldering condition, the resulting rush of air-oxygen into the room will produce an explosion of flames from hot gases that can blow out windows and blow doors off their hinges. Such an explosion is called a *backdraft*. Always feel the surface of any wall, door, or any metal doorknob to a closed-off room suspected of being on fire. If any of these are hot, the room is probably on fire or smoldering!

Now that we know a little about fire, it is time to get to the big question. What are some of the ways that firefighters, and maybe you or your family, can extinguish it? We will discuss this in our next chapter.

4 How Fire Can Be Extinguished

To Stop a Fire

Whenever anyone is faced with one of the more common types of fires, such as wood, grass, cloth, or paper, any one of the following fire-stopping actions can be employed:

1. *Prevent oxygen from reaching the burning fuel*
2. *Cool the heated fuel to slow and stop chemical reactions*
3. *Remove the fuel*
4. *Interfere with the chemical reactions of fire*

Most fires can be stopped by preventing oxygen from reaching a burning fuel. Small fires can be smothered by throwing blankets, tarpaulins, sand, or dirt over them. In cases of a grease fire in a kitchen frying pan, a simple method of smothering is to put a large cutting board over the flaming pan. Special bubbly foams can be used to blanket fires to prevent oxygen from getting to the burning surface. Water fog spread over the base of a fire can produce great quantities of water vapor or steam. Such an inert atmosphere can blanket a fire, reducing the amount of oxygen that can get to the fuel. It also cools the burning fuel and the area around it, which aids in the suppression of the fire.

Note that when water molecules are driven free from boiling water they form an invisible gas made up of billions of single water molecules. When these molecules cool somewhat they begin to cling together, or condense, forming a white cloud. The *steam* we see is actually hot water-gas condensed into small, boiling hot (212° F) water droplets. Water-gas in a sealed container may still be considered steam, but if under pressure it can now be heated to many hundreds of degrees hotter than water boiling on a stove.

When a fire is in a depression, it is sometimes possible to use carbon dioxide to put it out. Carbon dioxide, being about 50 percent heavier than air, may tend to lie down over such a fire and interfere with oxygen reaching the fuel. But this is not very often possible because of the violent air movement around a burning fire.

Most fires can be stopped by preventing oxygen from reaching the burning fuel.

A wood fire can be cooled by absorbing some of the energy being radiated from its overly heated and flaming outer wood molecules by drowning it in water. This is usually not possible. But a fire can be cooled by playing a water spray or fog from fire fighting nozzles on it. Cooling a fuel makes chemical reactions with air-oxygen less likely. Water absorbs infrared radiation or heat quite efficiently.

It is interesting that the amount of heat a drop of water must absorb to convert it from a boiling condition (212° F) to water-gas, or steam, which is also at 212° F, is almost 1,000 times more heat energy (actually 971 times) than is required to raise the same drop of liquid water 1°F, say from 200° to 201° F. Water has to absorb a tremendous amount of heat from a fire to change from its liquid form to its gaseous form. For this reason, water really absorbs radiant energy and cools a fire!

By breaking up water from a fire hose into very small droplets, the surface area of the water compared to its volume is greatly increased over that of a solid stream. Heat energy that is absorbed by a small drop of water will kick free surface molecules of the drop in all directions. These reduced-size droplets can then absorb even more efficiently any heat energy radiation in the area. Actually, water absorbs great quantities of heat very quickly when changing from a fine fog to a vapor. The result is a significant cooling effect in any area where a water fog is used.

On extensive and hot fires, water fog may be converted to vapor long before it can

approach the site of the fire itself. In doing this, it is accepting a great quantity of heat radiation from the direction of the fire. Such a fog can be used to protect firefighters from excessive infrared heat radiation and act as a protective cooling wall or shield. A second and more directive water spray from behind such a broad protective fog wall can be used to attack very hot fires. Fires cannot be brought under control and be extinguished unless they are cooled faster than they are chemically generating heat.

Extinguishing Fires

Assume a fire is radiating 10,000 btu (2,520,000 or 2.52×10^6 calories) from a vertical surface area, such as a wall, in a second. If a solid stream of water is played on it, part of the water will run off, and perhaps only 4,000 btu will be absorbed by the water. In this case, the fire's chemical chain reaction is not being broken, and the fire continues to expand. Should the same amount of water be applied in spray form over the same fire area, the small droplets of water, with their relatively greater surface area than is provided by the solid stream, could accept much more of the infrared energy radiation from the fire. The spray might effectively accept 10,000 btu of energy, and the fire might no longer grow in size.

Suppose the same amount of water is broken up by a high-pressure fog nozzle into a water fog in which the droplets are microscopic in size. The ability of the water in fog form to absorb heat radiation in a given time is greatly increased by the still greater water surface area of the microscopic drops over that of small spray droplets. Such a fog might be able to accept 15,000 btu of heat radiation in a second. Thus, 5,000 btu can be working to reduce the heat of the fire. If the fog is turned off too soon, the fire will rekindle due to outward radiation from still energized molecules deep below the fuel surface. It is necessary to continue to cool all areas of the fuel, the surface, beneath the surface, in cracks, etc., to a temperature below the ignition point of the fuel to break the chemical chain reaction of a fire.

One pound of liquid carbon dioxide (CO_2), usually considered a very good fire extinguishing substance, when loosed as a gas and snow-like flakes on a fire, has a cooling effect of about 150 btu. In comparison, 1 pound of water in fog form can accept heat energy equivalent to about 1,200 btu. This is a ratio of about 8 to 1 in favor of water fog over CO_2 as a coolant for Class A fires.

Measurement of Heat Energy

Heat energy can be measured in British Thermal Units, called *btu*. One btu is the heat energy needed to raise one pound of pure water 1° F.

Textbooks tell us that 1 btu = 1,060 *joules* of energy. Since 1 joule of any type of energy is equal to 1 watt/sec of electrical energy, then 1 btu = 1,060 W/s. To make this a little more meaningful, 1 btu is approximately the amount of heat given off by a 100-watt light bulb in 10.60 seconds. This is a lot of heat, as you probably know if you have ever touched a glowing 100-W incandescent lamp!

Heat can also be measured in *calories*. One calorie is the heat required to raise one gram (g) of water 1° C (Centigrade or Celsius), or 0.56° F. (If you are not acquainted with metric weights, a U.S. nickel weighs about 5 g.) Textbooks say that it takes 252 calories of heat to equal 1 btu.

Since fires involve great quantities of heat, the larger btu unit allows smaller numbers to be used in fire discussions. The smaller calorie unit is often used in chemistry experiments, for example. Maybe those of us who are on diets might prefer to talk in btu's instead of calories—ice cream having 252 calories would only have 1 btu! A normal daily dietary intake might be only 4 to 8 btu.

Another method of extinguishing a fire is to flood it with a dry powder that interferes with the chemical chain reaction that is occurring at the surface of the fuel. Either sodium bicarbonate (good old baking soda) or potassium bicarbonate can be used to produce this effect. When blown or thrown at the base of a Class A fire (or on flaming grease), it will interfere with the oxidation process and the flames will cease. If the heat is not too deeply seated, the fire will stay out. Special additives can be used to make these powders sticky when heated to enable them to adhere better to vertical or overhead surfaces when used in fire extinguishers.

Somewhat similarly, there are commercial substances that can be painted on wood to act as fire retardants. These fire retardant treatments turn the surface into something that acts like a charred area when it reaches a temperature of about 500° F. The charring prevents oxygen interaction with the wood below it, which can stop or greatly hinder the progress of fire. It is desirable for future homes to employ such treated woods.

Removal of fuel seems a rather obvious means of preventing the spread of fire. Unfortunately, it is not too often possible. It is accomplished in grass and wildland fires by digging or bulldozing a *fire line* down to *mineral earth* (soil without any combustible materials in it) across the path of the fire. Trees ahead of a fire can be felled away from the fire line. Sometimes *backfires* can be set up ahead of a fire by firefighters to develop a burned-out, and therefore a no-fuel, area to stop the progress of a moving grass or brush fire.

When fires develop on the surface of open tanks of oil or gasoline, it may be possible to drain off some of the liquid from the bottom of the tank through drain pipes, thereby starving the fire. Once liquid fuel is in a pipe, only the fuel at the open end can make contact with oxygen and support flames. Such a fuel line may be capped, and the flames smothered. However, if there is both air and fuel in a pipe, an explosion may occur if the pipe end is ignited.

Cooling a fire can also be accomplished by driving enough air across the fire surface to separate the flame from the fuel, at the same time air-cooling the fuel surface. This is what happens when you blow out a match. But this is not a very practical method of fire suppression in practice because a strong enough air stream can rarely be produced. However, it is possible to have sudden strong reverse winds suppress a grass fire line. A blast of dynamite near the base of certain types of oil, gas, or gasoline fires, as in the Kuwait oil fields in 1991 during the Persian Gulf war, can accomplish sufficient separation of fuel and heat to stop the flames. It was also proven in that war that large enough quantities of high-pressure water fog could do the same thing. A high velocity CO_2 gas cloud discharge can separate smaller flames from their fuel, substituting inert CO_2 gas for air-oxygen and cooling the fuel surface at the same time.

Stopping oil well fires takes a lot of ingenuity and experience.

Extinguishing With Water

Luckily, water is the best substance known to absorb heat and is reasonably available in most cases. But how much water is necessary to put out a fire in a structure? Can a reasonable answer be given to such a question? Let's see.

It has been established by tests that when a single pound of a normal combustible Class A material, such as wood or clothing, burns completely, a total of about 8,000 btu is developed. (Plastics, oils, and gasoline produce about twice as many btu per pound.) A dwelling can be considered as having roughly ten pounds of combustibles per square foot of floor area. Thus, a bedroom with dimensions of 12 feet x 13 feet, has about 150 sq. ft. of floor area and therefore about 1,500 pounds of combustibles (floors, walls, ceiling, and furnishings). At 8,000 btu per pound this represents a total release of 1,500 x 8,000, or 12,000,000 (1.2 x 10^7) btu if completely burned. Assume the fire is being attacked within some reasonably short period of time after it started. Perhaps

only 120,000 btu is being liberated by the fire per minute. How much water is required to cool this much heat energy?

To make 1 lb. of water pass from a boiling liquid state at 212° F (100° C) to steam (also at 212° F) requires almost 1,000 btu. To cool an area generating approximately 120,000 btu in a given time would require 120,000÷1,000, or 120 pounds of water. Since water weighs a little more than eight pounds (actually 8.334) per gallon, it would be necessary to release rapidly, about 120÷8, or about fifteen gallons of water in fine fog form into the room to cool it to the point where the fire no longer is increasing. This does not appear to be much water to quench a room that is well involved in fire, but what this amount of water would do is to "knock down" the fire. It would still require a continuing, although possibly reduced, attack to be maintained to keep the walls and ceiling cooled enough not to re-ignite or continue to burn. If the attack were started only one minute later, the fire might have doubled the amount of heat developed, and more than twice the water would be needed.

The rate at which water is applied to a fire affects the amount of water required. From tests made when extinguishing fires, the application of the required amount of water coolant in about thirty seconds seems to be most efficient. For the fire in the explanation above, in which it was estimated that a total of about fifteen gallons of water in fog form would be needed, attacking it for thirty seconds with a fog nozzle rated at thirty gallons per minute (gpm) should knock it down to a relatively easy extinguishing and mop-up condition.

If the fire had been allowed to burn only a minute or so longer, thirty seconds with a sixty gpm fog nozzle might be required to knock the fire down. After only a few minutes more, it might have required a 100 gpm fog application, and so on. (It should be understood that our examples are about smaller structures and just to illustrate some very basic ideas.) When a building is greatly involved in fire, the firefighters may have to stay back and pour as much water as is available into the fire area.

Would it have been possible to use the 100 gpm fog application in the first case when only fifteen gallons were needed? Of course, and it would have required quite a few seconds less time to knock down the fire. It might not have been quite as efficient, hydraulically speaking, but it might have prevented the damage that could have occurred during those several seconds of burning that were never allowed to occur. However, there would probably have been much greater water damage to the contents of the room. It may be important to consider that for every 100 extra gallons of water trapped on a floor, there will be an additional weight load of 100 x 8.33, or 833 extra pounds, on a possibly weakened floor. Only considerable fire training enables firefighters to figure out the most efficient method of attacking fires.

A simple formula that firefighters may use to approximate the gpm of water needed from a fog nozzle to cool a given volume room is:

$$gpm = cu.ft. \div 100$$

If a small bedroom is 12 ft. x 13 ft. x 8 ft., its volume is 1,248 cu.ft. This, divided by 100, is about 12.5 gpm, or about 25 gallons in half a minute.

These theories indicate that anyone fighting a fire should mount a strong initial attack. Firefighters may want to shut down their nozzles' output as soon as flames are knocked down to prevent excessive water damage to a room.

Extinguishment By Foams

One of the important fire extinguishing means, particularly useful on flammable liquid or oil fires, is smothering to prevent oxygen from reaching a burning fuel. This may take the form of large blanketing sheets, such as tarpaulins, or covering the fire with tough, small-bubble foam. There are two basic types of liquid foams that may be found in use, chemical and mechanical.

A chemical foam is developed by mixing inorganic chemicals (chemicals not derived from animal or plant life) with water to produce a smothering blanket of small bubbles containing carbon dioxide. Chemical foams can be low expansion types, producing about ten times the volume of foam as was present in the chemicals, plus the liquid. A total of ten gallons of water and chemicals can generate about a hundred gallons of smothering foam.

A mechanical foam can be produced by mechanically mixing a protein-based, or organic, foam-generating powder, with water, and then introducing this solution into the water stream of a hose line. This is accomplished with a piece of equipment known as an *eductor*. Fig. 4-1 illustrates how such an eductor, or proportioner, operates. As water under pressure flows along the hose line from left to right, it passes into a reduced-opening section that increases its velocity. High velocity water passing the small *Venturi tube* opening, suddenly expands into a more open area which will therefore have less pressure. The area of reduced pressure, or suction, pulls any solution inside the Venturi tube out into the expanding water stream. In this way, the foaming agent is mixed into the hose water. As the water and foaming solution passes through the fire hose nozzle, it mixes with air, producing billions of tiny bubbles that are expelled from the end of the nozzle as a foamy spray.

High expansion foam solutions can produce foams having many times the volume of the water used. The first foaming agent used was a 6 percent solution. With such a solution, 6 gallons of foaming agent added to 94 gallons of water produced 10 x (6+94), or about 1,000 gallons of somewhat sticky, smothering foam. With a more concentrated 3 percent solution, 3 gallons of foaming agent educted into 97 gallons of water produce about the same 1,000 gallons of foam. Different concentration solutions require different eductors, unless there is a calibrated control valve on the eductor foam line to set the amount of foam that can be educted.

Figure 4-1 shows a foaming agent shut-off valve just ahead of the Venturi opening. Foams are sometimes laid down ahead of some types of advancing fires to slow or stop their advance.

In some fire engines, the foam eductor system may be built into the water outlet system to one of the hoses. There are also small 2½-gallon portable extinguishers that can generate about 25 gallons of foam in about one minute.

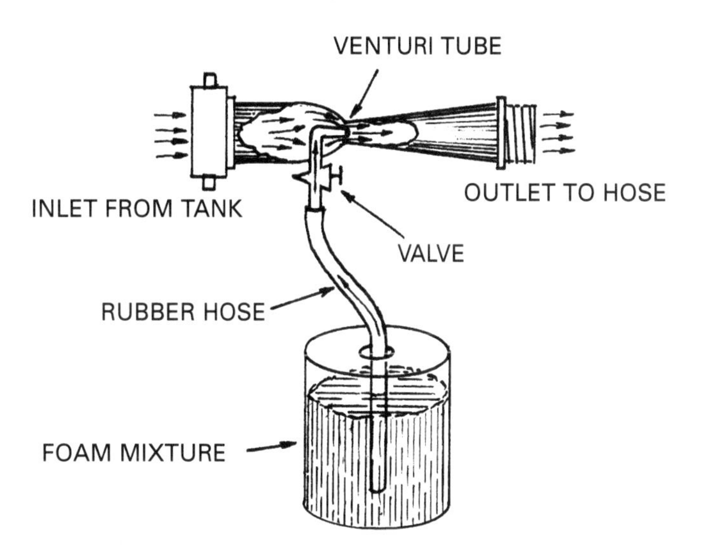

Figure 4-1. Eductor idea to inject foam mixture into a fire hose line.

Eductors are also used to mix liquid *wetting agents* into hose lines. Normally, water tends to form drops on surfaces. A wetting agent, such as soap, allows the drops to spread out over a surface and down into pores and cracks in the surface material. This greatly increases the effectiveness of water in fighting fires.

Foam blankets are particularly effective on open oil or gasoline fires. However, when fed across an open flaming tank, the foam must be directed against the inside of the tank wall to allow the foam to spread out without great violence across the flaming surface. If a high-velocity foam stream is aimed at the surface of a flaming tank of oil, the result may be a spraying outward of burning oil and a drastic increase in fire spread.

The general purpose protein-based foams are not effective on flaming alcohols, acetones, ketones, ethers, or various petroleum products. Such substances may combine with

Small ordinary combustible material fires can be extinguished by using several types of commonly available fire extinguishers.

the foaming agent and break down the bubbles. Special "all purpose" foaming agents, as their name indicates, can be used on all types of flammable liquids as well as on normal combustibles.

High-expansion (1,000 to 1) foam generators can produce enough foam to completely fill a relatively large room in a few seconds. Their electric fans blow a mixture of foaming solution and water through a fine wire mesh to generate the billowy foam. These devices are useful for fires in basements, bowling alleys, warehouses, or other large-volume areas.

Primarily useful against Class A and flammable liquid fires, foams should not be used on metal fires. Also, and this is very important to the firefighter: Since all foams are electrical conductors, they should not be used on fire areas where they might be sprayed on any electrical circuits that are still alive.

Small Class A fires can be extinguished by using several types of commonly available fire extinguishers (Chapter 5), but water in fog, spray, or solid streams are necessary in larger fires.

Classes of Fires

So far we have identified only one class of fire, basically the wood or cellulose, specified previously as Class A. However, there are actually four classifications of fires:

1. *Class A—Ordinary combustible materials*
2. *Class B—Flammable liquids or greases*
3. *Class C—Fires in live electrical circuits*
4. *Class D—Fires due to chemical reactions other than oxidation*

The Class B grease fires can be contained by smothering with foam or some kind of blanket or tarpaulin, by dousing with dry powders, and by applying properly used carbon dioxide extinguishers. Water in stream form may only spread Class B fires. Most of the flammable liquids, such as oils, have a low *specific gravity* (*sp. gr.* is the oil's weight per unit volume compared to water's weight). If a water stream is applied to burning liquids in a tank, the water, having the greater sp. gr., sinks below the flaming liquid and is then completely ineffective in extinguishing the fire. Water fog, on the other hand, if dense enough and if applied just above a burning oil surface, may exclude enough oxygen and cool the oil surface enough to stop the chemical reaction of burning.

Closely allied to flammable liquid fires are gas fires, such as occur with broken gas lines. While Class B techniques of extinguishing such flames might work, often it may be better to allow the gas flame to continue to burn, but concentrate on keeping nearby areas cooled below their ignition temperatures until the gas main is turned off, or until the gas runs out. If a gas flame at the end of a broken gas pipe is extinguished, the raw gas will continue to be ejected and is very dangerous because it is explosive and poisonous.

In a home gas fire, the first thought should be to turn the gas off at the device, or at the inlet valve outside, or at the storage tank, if out in the country. Notify the fire department first, then try to work on the fire afterwards.

Electrical fires are only Class C fires while the electrical circuit is alive. Once the power lines are dead, fires started by electrical sparking and heating will usually have to be fought as Class A fires. While the lines are alive, however, CO_2 gas is probably the best extinguisher to use on electrical fires, with dry powders next. Both of these are considered electrical insulators or non-conductors.

In most electrical structure fires, one of the first things to do is to open the main switch at the electric meter box to disconnect all electricity in the building. At night, if the lights are on in the building, it will be up to the fire chief or incident commander to decide on the proper action. It may be easier to fight a fire in lighted areas, however.

It is difficult to generalize about Class D fires. As mentioned before, active metals such

as sodium, potassium, and lithium (which is used in some modern batteries), will violently react with water so that such fires cannot be extinguished by water. Usually, CO_2, which is an inert gas, does not combine with Class D fire metals, but its cooling effect is limited. Special powders are available with the consistency of fine sand, made of organic salts and graphite that can be used on Class D fires. They turn into an inert paste-like liquid when heated and can interfere with the chemical burning action. Portable Class D fire extinguishers use such powders. An example of an active-metal fire might be a burning magnesium automobile engine. As mentioned before, active-metal fires may also be smothered by piling dirt or sand on them.

It is important to know about the many kinds of fire extinguishers, how they work, and for what classes of fires they are designed. Some of the extinguishers carried on fire engines or designed for use in home or shop are described in our next chapter.

5 Portable Fire Extinguishers

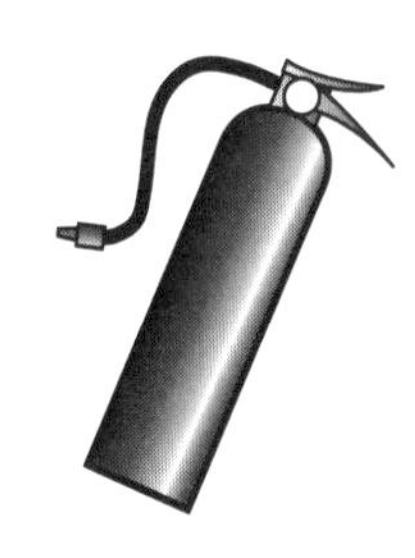

Putting Out Small Fires

If you are present when a small fire starts, you will have to do something in a hurry, or much damage may result. Depending on the fire, think of these possibilities—smothering it with a blanket, dumping water on it, beating it out with a wet sack, using a portable fire extinguisher on it. All of these methods can be used and have resulted in the extinguishing of millions of small, but potentially disastrous, fires. But in many cases, a fire extinguisher is the best tool for putting out a small fire.

There is a wide variety of portable fire extinguishers. Those carried on fire engines are 2½-gallon or larger sizes. For the automobile, home, shop, or kitchen, smaller 2- or 2½-lb size dry powder extinguishers should be adequate first aid for small fires. Fire engines may use 2½- to 30-lb tank extinguishers, whereas industrial wheel-around tanks may range from 150- to 350-lb tanks of dry chemicals.

If you find it necessary to fight a fire in a home or other building and find that the fire is not coming under control within about five seconds of your using an extinguisher, waste no time in calling 911 or your local fire department. A 911 system usually shows the dispatcher the address of the telephone being used and its subscriber's name. You may only have to say that there is a fire, where it is, how fast it is spreading, and your name as the reporting party. Then make sure the building is cleared of all occupants. After that, you might go back with the fire extinguisher to see if you can knock down some of the fire or remove burnable objects from the fire scene. If you cannot get back into the room on fire, close it off, if possible. If you can, close all the windows and doors of any room in which there is a fire, as well as all of the other rooms of the house which are easily accessible.

If it is daylight, shut off the main electric switch at the meter. You might also try to spray water onto the base of the fire with a garden hose until the fire department arrives if there are not hot electrical circuits involved. Beware of inhaling smoke—it may be poisonous!

Fire Extinguisher Ratings

If your fire extinguisher does not knock down a fire in about 5 seconds, something else must be done in a hurry!

Modern fire extinguishers, which are all passed by the Underwriters Laboratories (UL), will have a rating on their name plates. The rating consists of one or more numbers and letters. The letter indicates the class of fire on which that particular type of extinguisher is effective. The number is an indicator of the duration of its effectiveness on such a fire. Fire extinguishers may carry ratings for Classes A, B, C, and perhaps D fires (fire classification descriptions are given in Chapter 4).

An extinguisher carrying a 1A rating can be expected to put out a Class A wood type fire involving 50 sq. ft. of wall area (such as an eight-foot high wooden wall about six feet wide). The 2A rating for a standard 2½-gallon water stream extinguisher indicates it should be able to extinguish about twice as much wall surface as a 1A rated extinguisher would. These are only approximations because trained firefighters will use extinguishers more efficiently than inexperienced users. A trained person would normally start applying the water or other extinguishing agent near the base of the flames. Most untrained people might aim at the top or middle of the flames.

Extinguishers for Class B fires (flammable liquids, such as paint, grease, oils, gasoline, etc.) are rated by the number of square feet of an open tank surface of a gasoline fire that the extinguishing agent can suppress. It is significant to note that a 10-lb carbon-dioxide (CO_2)

extinguisher may be rated 8B:C, whereas a 20-lb CO_2 extinguisher is rated as only 12B:C. The larger the burning surface, the more difficult

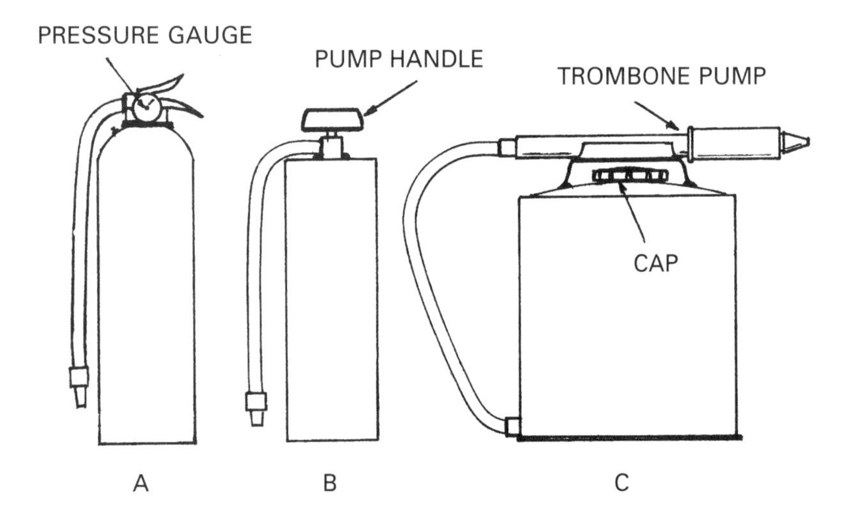

Figure 5-1. Pressure and hand-pump extinguishers, (a) water or foam, (b) water or loaded stream, (c) forestry.

the fire is to suppress, and the relatively greater amount of extinguishing agent required. Note also that this particular extinguisher carries a Class C rating, indicating that it is an electrical nonconductor and therefore can be used on electrical fires. No quantitative Class C rating is given because electric fires will be confined to some point, i.e. a motor, an electric switch box at crossed or overheated wires, etc.

One of the simplest of the portable fire extinguishers is the pressurized water type (Fig. 5-1a). To charge it, the cap is unscrewed and the 2½-gallon metal tank is 90 percent filled with fresh water. The cap is then replaced, an external air pump is coupled to the air chuck or filler fitting, and the internal pressure is brought up to about 100 pounds per square inch (psi). Pressure is shown by an air-pressure gauge mounted next to the trigger on the handle. If you want to use this extinguisher on a fire, you must pull the safety locking pin from the valve assembly at the top of the tank. Then by pulling back on the valve trigger, a stream of water is propelled out the hose and nozzle. A 3/16-inch stream of water can be thrown about 20 feet. Remember, playing the stream at the base of the fire will probably be most effective. Holding a thumb partially over the nozzle produces a spray which may be more effective than a solid stream if it is possible to make a close approach to the fire. These extinguishers have a 2A rating and should be checked to be sure they are always kept at a pressure of 100 psi or more.

Such fresh-water-stream extinguishers can be made more efficient if a wetting or penetrating agent is added to the water. If they are expected to be used in subfreezing temperatures, noncombustible antifreezes such as keloy or potassium carbonate might be added to them. With such additions they are called *loaded stream extinguishers.*

Water normally has a *surface tension* which tends to make it form into small balls or droplets on any slick surface. The tendency to ball up prevents easy penetration of water into the fibers or cracks in wood. Depending on the type of wetting agent used, the addition of 0.2 percent to 0.01 percent of such agents (which are similar to soaps or laundry detergents) will relieve the surface tension. The water will not form droplets but will flow evenly over the whole surface and into cracks. Wetting agents can be added to the water in the tank of fire engines. They have no harmful effect on tank linings or hoses and may be educted into a fire hose line, as explained in Chapter 4. Water with a wetting agent in it may foam slightly when coming out of a nozzle, but it is not enough to produce a fire-fighting foam.

There are several hand-pumped fresh-water fire extinguishers. One, Fig. 5-1b, is a 2½-gallon cylinder with a double-action pump operated by pushing down and pulling up on the pump handle. With the extinguisher on the floor or ground, the handle is pumped with one hand, and the hose is directed at the fire with the other. A usable stream can be thrown a distance of more than 30 feet. The device has a 2A rating but may be much more effective if a wetting agent is added to the water.

Another hand-pumped water extinguisher is the 5-gallon backpack forestry type, Fig. 5-

1c. These are included as necessary equipment on most grass and brush fire fighting vehicles. It is fitted with two snap-on carrying straps (not shown) that allow it to be carried on the firefighter's back. The barrel is held in one hand, and the pumping action is produced by sliding the trombone-like piston pump handle out and back with the other hand. Its two-choice nozzle may be changed to produce a straight stream or a spray. It can also be operated resting on the ground. Since they contain five gallons of water, they carry a 4A rating. Adding wetting agents increases their efficiency. Their tanks may be galvanized iron, stainless steel, or fiberglass.

A newer form of forestry backpack extinguisher is the *bladder pack*, a 7-gallon collapsible rubber bladder with a piston pump attached to it. It is used in a manner similar to the 5-gallon rigid-tank forestry extinguisher.

Gas-Type Extinguishers

Carbon dioxide, or CO_2, gas extinguishers consist of heavy-walled (3,000 psi test), rounded steel cylinders or tanks (Fig 5-2b). When fully charged, such tanks hold from 5 to 20 pounds of liquid CO_2, depending on their size. When charged, they may weigh from 20 to 60 pounds. The gas is under about 900 psi pressure in the cylinder. At this pressure it is a liquid. To operate the device you must pull out the safety locking pin from the handle, point the horn-like nozzle (attached to the end of a thick black insulating hose) toward the base of the fire and pull the trigger. A heavy discharge of CO_2 gas is blown out of the nozzle a distance of twenty feet or more. It appears as a white snow from the moisture of the air that it freezes. However, to be most effective, the discharge should be applied from a distance of five to ten feet if possible. Since the gas expands greatly as it is released, the outermost electrons of the molecules of the gas immediately lose energy due to the lack of pressure on them, and the CO_2 becomes so cold (accepts heat energy so readily from the air) that it will cause frostbite if allowed to strike human skin. Live electric-circuit fires can be safely cooled and hopefully smothered with a stream of CO_2 gas.

The best way to determine how much charge is left in a CO_2 extinguisher is to weigh it. A 5-lb CO_2 charge plus a 15-lb tank produces a 20-lb extinguisher. If it has been partially discharged and weighs 17½ lbs. (lost 2½ lbs.), it must be half depleted. To be considered still usably charged, a 20-lb extinguisher should weigh at least 95 percent of its weight when full, or 20 x 0.95 = 19 lbs.

CO_2 extinguishers should be checked for pressure at least twice a year even if not used.

CO_2 extinguishers should be checked at least twice a year even if not used. The steel tanks must be tested at least once every five years to make sure that they are still in a safe condition to store the gas at high pressure.

In the 1980s, Halon, a bromine gas, appeared as an extinguishing agent. One squirt of it at a kitchen stove grease fire could put it out immediately. Halon interferes with the chemical action of most fires. It was available in portable extinguishers and ceiling-mounted sprayers. Unfortunately, bromine gas poses a threat to the ozone layer of the atmosphere and was banned after 1993, although some charged extinguishers may still be around.

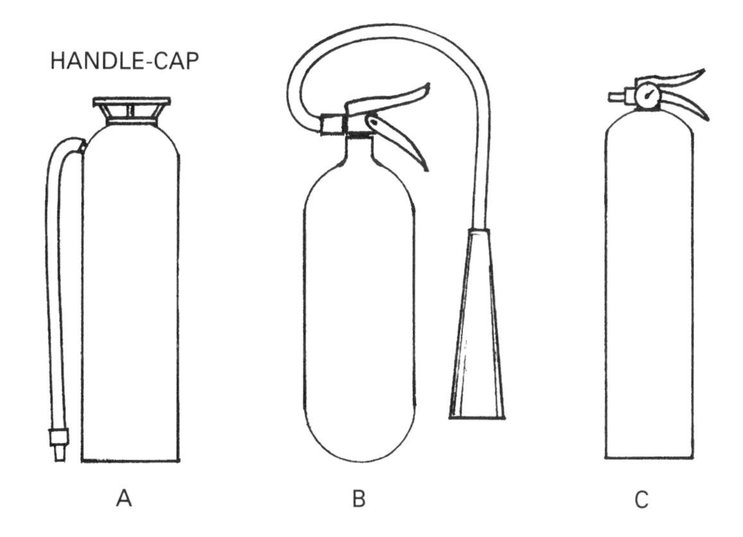

Figure 5-2. Extinguishers. (a) Soda-acid, (b) Carbon-dioxide, (c) Dry chemical.

Dry Powder Extinguishers

The most popular extinguisher over the years has been the dry chemical pressurized kind, Fig. 5-2c. The powder in it is a specially treated white sodium bicarbonate. Purple potassium bicarbonate powders are also available, which are somewhat more effective than the sodium bicarbonate. These extinguishers are nearly filled with a noncorrosive, nonconductive, and non-freezing powder under 100± psi pressure from an inert gas, such as nitrogen. It is necessary only to pull out the safety locking pin at the top, aim the nozzle at the base of a fire, and squeeze the trigger. A cloud of powder will fly out about ten feet from the nozzle. When it hits the hot surface at the base of the fire, it chemically changes (through what is known as *free radical combination*) to CO_2 gas and water. These interrupt the chemical chain reaction of the fire, smothering, cooling, and chemically stopping the flames. Some powders have additives to give them good clinging capabilities, making them effective on hot walls and ceilings, as well as on Class B and C fires. All these extinguishers have small pressure gauges on them, which indicate if the internal pressure is adequate (green area). If the pressure drops (red area), the unit should be discarded, although some can be recharged at local fire equipment outlets. To discard, trigger until empty, then dispose of it in a metal recycling dump.

Mount an extinguisher in your kitchen near the stove, on every floor, in your shop, and in each vehicle. Be sure all persons in your home are taught the proper use of extinguishers.

There are several types of triggering systems that may be found. When the pin is pulled out:

1. *The trigger may be a small push-in button type.*
2. *Releasing the pin may free a vertical handle, allowing a push-down trigger to eject the powder.*
3. *In newer models, squeezing a pistol-grip type of handle will expel the powder.*

One kitchen extinguisher is an unrefillable, small-bubble CO_2-foam type. If applied to the base of a small fire, the foam flows over the burning area, extinguishing the flames, leaving a thin non-ignitable coating.

Special multi-purpose "ABC" powders contain ammonium phosphate instead of bicarbonates. These are particularly recommended for small Class A fires, although all dry powders are useful to some degree on this class of fire. The pressure gauge above the trigger-handle of this kind of extinguisher indicates its state of charge.

Depending on the type of powders used, some extinguishers having ten pounds of powder may carry ratings of 20B:C, 40B:C, or 2A:30B:C. The usual home, automobile, shop, or kitchen 2- or 2½-lb type extinguishers are sometimes rechargeable and may carry a 5B:C, or 6B:C rating. Note that these are only rated for grease, oil, and electrical fires. They are not very effective on wood and other Class A fires.

An *outside-cartridge* type of dry chemical extinguisher has a replaceable high-pressure gas cartridge plugged into the side of its powder filled tank, near the top. To operate it, the safety pin is pulled out, the cartridge is punctured by pushing down on a lever above it, putting the powder in the tank under about 600 psi pressure. It is possible to throw a cloud of this dry chemical more than 20 feet, but the most effective range is closer to half of this distance. A 10-lb capacity tank may carry a rating of 16B:C, 40B:C, or 2A:20B:C, depending on the powder with which it is charged. Once punctured, the cartridge loses its pressure, and the extinguisher must be recharged with powder, and a new gas cartridge installed. This can be done in the field in a couple of minutes. Recharging of most rechargeable extinguishers is done at local fire equipment companies.

A popular fire extinguisher used in the past was the *Pyrene* (carbontetrachloride, or CTC) type. It came in pint or quart size brass bottles filled with liquid CTC. When the pump handle on the top of the bottle was worked in

and out, a thin stream of CTC was ejected from a nozzle at the bottom. Unfortunately, the fumes of CTC are dangerous to health if they are strong enough to smell. (CTC was used to gas moles and gophers!) If you happen to find one of these extinguishers on a wall somewhere, be careful if you must use it! They did work quite well on many small fires.

Active-metal fires can be extinguished only with special dry powders or by burying them under sand or dirt. Extinguishers using the required powders come in 10- to 30-pound sizes, are carried on fire engines, and may be found in many industrial plants.

Mount an extinguisher in your kitchen near the stove. You should have at least one extinguisher on every floor of your home, in your shop, and one extinguisher in each of your vehicles. Be sure all persons in your home are taught the proper use of all types of extinguishers on hand.

Portable extinguishers may be able to extinguish small fires of various types for you, but when a fire continues to increase in size while under a fire extinguisher attack, the only hope of putting it out is probably by using large quantities of water. This usually means fire engines and firefighters, so call 911! Let's take a look in our next chapter, at the make-up of a basic fire engine, and how it might be operated to provide large quantities of water to fight fires with fogs, sprays, and straight streams from hoses.

6 Fire Engines and Their Pumps

The Fire Engine

The workhorse of any fire department is known as its *pumper*, *fire engine*, or simply *engine*. The basic pumper will usually have a self-contained water tank with a 300 to more than 1,000 gallons water capacity, and one or two water pumps. The amount of hose it will carry may vary. A possible load might be: 1,000 feet or more of 2½-inch or larger hose; perhaps 400 feet of 1½-inch hand line; 500 feet of supply line (3- to 5-inch hose); one or two 200 foot reels of always-charged, ¾-inch or 1-inch high-pressure *hard-line*, also called *booster* or *red line* hose.

Besides the basic water, hoses, and pumping equipment in a fire engine, it will have an assortment of hose fittings, wrenches for these fittings and other tools, ladders, several types of fire extinguishers, forcible entry tools, pike poles, salvage covers, first aid kits, breathing apparatus, and two-way radio equipment.

For mountain area rural departments, the wheelbase of fire engines might not exceed twelve feet in order to allow them to navigate safely on sharp mountain turns. While four-wheel drive might be an advantage, pumpers are not normally expected to operate off county roads, private roads, or highways. They usually park on the nearest hard ground and lay out their hoses to the scene of the fire. Radio communication with persons at the fire scene is used regarding water requirements. For city street operation, the wheelbase of an engine may be sixteen feet or longer.

Particularly in rural situations, truck weights may be an important consideration if private wooden bridges have to be crossed or if roads become muddy in rainy weather. Since water weighs 8.33 lbs per gallon, the water alone in a 500-gallon tank weighs over two tons. If the pumper and its other fittings weighs another three tons, which would be a light unit, this represents a load of five tons, with perhaps three of the tons over the rear wheels. There are many home-built bridges leading to farms or homes, and possibly some older county road bridges, that might not stand a moving, bouncing load of this weight. It is imperative that all rural fire engine drivers find out how much load all of the bridges in their district can carry and drive slowly over weaker ones. Drivers must also know the width of all streets or roadways and height of branches over roads they might have to use.

The plumbing in all except the simplest engines allows water to be pumped at pressures ranging from 100 to perhaps 400 psi, and may include as many as five hose outlets mounted on the sides and rear of the pumper. Water can be pumped from the tank to these hose outlets. One of the rear outlets will usually have a *preconnected line* coupled to it. This allows one of the firefighters to jump out and haul to the fire scene the ±150 feet of preconnected 1½-inch hose which will normally be fitted with a combination fog/solid-stream/off nozzle. With this, water can be sprayed on the fire as soon as the engineer/driver can step out, start the water pump, and open the valve to the preconnected line.

The engineer's control panel, consisting of valves, gauges, and controls, may be located outside on the drivers side, just behind the cab. Some engines may have the control panel operated from a platform directly behind the cab. The platform is raised to a height about five feet above ground level to provide the engineer with good 360° visibility of the fire scene and surroundings.

Pumper Operation

Water is usually available from hydrants in cities, but most pumpers can also draft from

any pool of water. Such drafted water can be pumped to the engine's water tank, or directly to its hose outlets, or to both simultaneously. Water may be relayed through its pumps to another pumper several hundred feet or yards closer to the fire scene if the fire is a long distance from a water source.

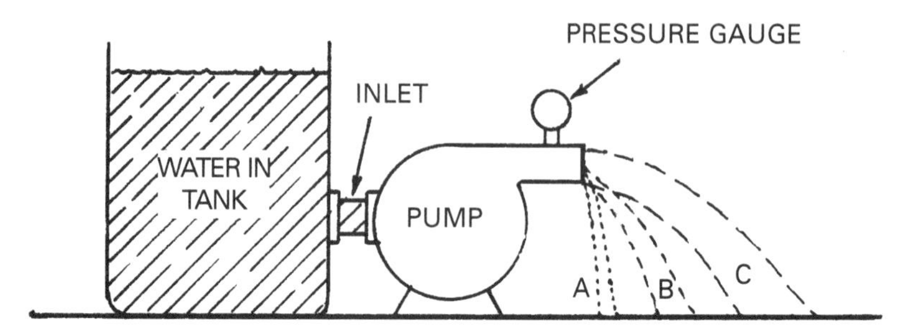

Figure 6-1. Pump-flows with no back pressure.

The stream of water directed at a fire is the result of the water pump on the fire engine.

City fire engines usually couple to a hydrant installed a couple of feet off of the street pavement. Hydrant pressure from either gravity flow from water tanks on hills, or water-works pumps, will force water into the fire engines at perhaps 40 to 80 psi through any 4- to 6-inch soft hose lines that firefighters couple between the hydrant and the fire engine.

The pump in the fire engine increases the water pressure to its fire line hoses, or to overcome gravity if a hose is laid on an uphill slope, or to force water up special fire line pipes that are built into high-rise structures. Such pipes may terminate at each floor of the building with one or more water outlets. Pumpers couple to street-level inlets on the front or side of the buildings.

Regulation of Water Volume and Pressure

Because of the complexity of a modern pumper water system, well-trained operators with a good basic knowledge of hydraulics and mechanics are necessary to assure proper functioning of the pumping system at a fire, when time is at a premium.

The stream of water directed at a fire is the result of the action of the water pump on the fire engine. The output from the pump is considered in two terms, volume in gallons per minute (gpm), and *pressure* in pounds per square inch (psi).

Consider water to be pumping directly out of a tank and into the air by the centrifugal pump in Fig 6-1. Since there is nothing except air against which the pump is working, the pressure of the water at the pump outlet is almost the same as at the inlet. Actually, there is a small positive pressure given to the water at the outlet, and a small negative pressure (suction) developed at the inlet. With nothing opposing the water flow, the number of gpm for this pump could be quite high. If the pump is rotating slowly it might produce only a dribble of water as indicated by stream "A". If the pump is driven faster, the velocity and energy given to the water by the pump will eject the water farther outward, as in stream "B". Still higher pump speeds will transfer more velocity and energy to the water, producing the stream shown as "C". With greater pump speed, a greater volume of water will be pumped. But the outlet pressure in psi will not increase materially, and the pressure gauge will still read nearly zero. This is a case of almost no outlet pressure, but a high output volume.

The system in Fig. 6-2 is similar, but has a short length of hose with a valved nozzle added to it. If the nozzle valve is closed (pushed forward), the pump will keep increasing the pressure of the water in the hose until either the hose bursts or the pump reaches its maximum possible pressure at the speed at which it is being rotated. This might be as much as 600 psi. The pressure differential between outlet and inlet is now the full 600 psi and would read as such on the pressure gauge.

If the hose is to be operated at 100 psi or less, it will be necessary to have some type of relief valve system across the pump, indicated

by the dashed lines. If you opened this relief valve a bit, high pressure in the hose would force water up through the valve and back to the inlet. Now, with the same pump speed, the hose pressure cannot exceed some lower maximum pressure. If you were to open the relief valve completely, the difference in pressure between pump inlet and outlet might be very small, and the hose pressure might drop to only a few psi. Such a relief valve could be manually controlled, but it is usually an automatic type having a spring-loaded valve. When the pump pressure increases to the point that the spring-loaded valve is forced open, the pressure can no longer increase in the hose line. For a given spring tension there will be some constant pressure that will be maintained in the hose line, 100 psi, for example. Besides regulating the pressure to the hose, the relief valve is also a safety feature to protect the pump and the hose.

If the nozzle valve is now cracked open a little, the 100 psi in the hose forces a stream of water out the nozzle. The hose and nozzle pressure is 100 psi, but the volume of water from the nozzle may be only a few gallons per minute if the nozzle opening is small. The larger the nozzle orifice, the more gpm discharged. Should the nozzle orifice be too large, the volume of water may be great, but the pressure of the water in the system might decrease below the relief valve setting, perhaps to only a few psi. The stream from the nozzle would be large in volume, but the *reach* of the stream would be limited so much that water might not reach distant flames.

To maintain enough pressure to throw a usable stream of water an effective distance from a nozzle, it is necessary that the nozzle orifice be smaller than the inside diameter of the hose being used. An example of an efficient nozzle orifice to hose diameter would be a ¾-inch nozzle on a 1½-inch ID hose. This 1:2 ratio of nozzle opening to hose diameter is a good working approximation. As long as the constriction by the nozzle orifice is adequate, the pressure in the hose line can be maintained. If the nozzle opening is small enough, the automatic relief valve, or *pressure regulator*, is forced to stay open by the internal back-pressure of the line. This maintains the desired maximum hose pressure, perhaps the 100 psi above. A stream with reasonably long reach should result.

You can infer from the above explanations that there must be some interrelationship of volume, pressure, and system resistance. It can be said that the water **Volume** delivered is directly proportional to the **Pressure** and inversely proportional to the **Resistance** in the system, or

$$V = P \div R$$

The resistance to the flow of water in a hose is mechanical. The molecules of water lying against the hose lining tend to attach

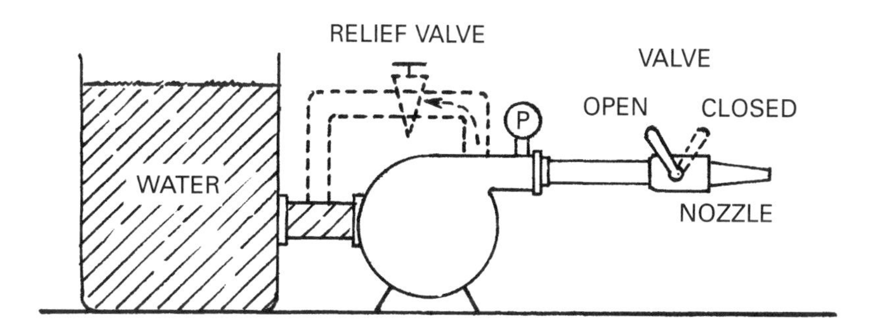

Figure 6-2. Pump system having hose, valve, and nozzle.

themselves to the lining, forming a relatively non-moving layer of water. The next layer of water molecules, nearer the center of the hose, will rub against the non-moving lining molecules. While these may move to some degree, they will not move as much as the next layer of molecules closer to the center of the hose, and so on. If the hose diameter is small, the resistance of the water molecules to movement (known as *stiction*) is relatively greater than it would be in a larger diameter hose.

Resistance due to small hose sizes means low water volume. Also, the longer the hose, the more the resistance, and the less volume

from a nozzle, assuming the same pressure at the pump outlet.

Another factor affects resistance. The greater the water velocity, the greater the turbulence developed in the water as eddy currents (little whirlpools) are formed due to wall surface irregularities, bends, and hose fitting discontinuities or constrictions. These all add up to a total resistance to water flow in a hose. With larger hoses, the lining resistance does not extend out toward the center of the hose as far (less stiction), and the resistance to water flow with the same pump pressure will be greatly decreased, although some turbulence might still be present.

For long hose lays, it is desirable to use as large a diameter hose as possible from the pumper to the fire ground, and then branch out with wye (Y) fittings to two or more smaller, more easily handled 1½-inch hoses, as an example, close to the fire. In this way, the greatest volume and pressure can be delivered to a distant fire with the most mobility of lines at the fire scene.

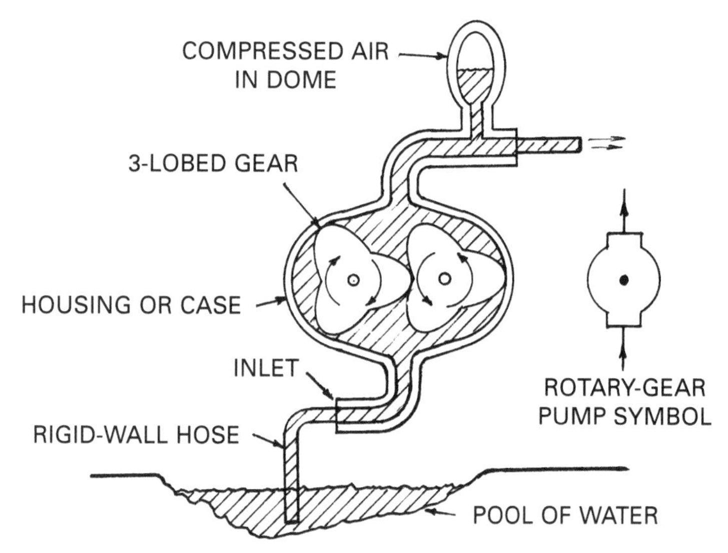

Figure 6-3. Basic 3-lobed rotary-gear pump and its symbol used in hydraulic diagrams.

It is interesting to consider the result of friction in rubber-lined fabric hoses of various inside diameters as far as loss of pressure is concerned using a 100-foot length of hose. With a 25 gpm water pump, a ¾-inch hose would lose 75 psi per 100 feet; a 1-inch hose would lose only 20 psi; a 1½-inch hose would lose merely 3 psi. With 2½-inch, 3-inch, and 4-inch hoses, the losses would be almost negligible. However, with a 150 gpm water pump as the source, the 1½-inch hose would lose 65 psi; the 2½-inch hose would lose 6 psi; and the larger hoses relatively less. For a 250 gpm source, the pressure loss of a 2½-inch hose would be 15 psi; a 3-inch hose would be 5 psi; a 4-inch hose would lose only 2 psi. It would be possible to carry water at 250 gpm for 1,000 feet and lose only 20 psi. Starting with 120 psi pressure at the pump, the pressure 1,000 feet away would be still 100 psi. For this reason, running four- to six-inch hoses from a city hydrant to a fire ground is an efficient means of transporting water and may be called surface water mains. (Fire hydrants in cities are fed by metal, or other material, below-surface water main pipes.)

Special water additives, variously termed slippery water, quick water, etc., have been developed. They coat the inside of hoses carrying them with an almost sticktionless surface of long, straight molecules that reduce lining resistance to water flow to values considerably less than those given above. These can be added to tank water or be educted into the hoses at the fire engine.

When the valve on the nozzle of a fire hose is shut down, it must be done so relatively slowly. Turning off the water rapidly reflects a shock wave back up the hose. The resulting water hammer can rupture hoses, break pump casings, and even damage appliances connected to the water mains downstream.

Types of Pumps

We have been discussing pumps in very general terms so far. There are three basic types of pumps employed in fire engines. The main pump is a *centrifugal* pump. If there is a second pump on an engine, it will usually be a *rotary-gear* or *rotary-lobe* pump. The third possibility, and least used except for some small, high-pressure applications, is the *piston* pump, similar in operation to the original fire engine pumps explained in Chapter 1.

Rotary Gear Pumps

The pump shown in Fig. 6-3 is one of several similar rotary-gear pumps. If either one of the two gears is driven in the direction shown by the arrows, the other gear is forced to turn in the opposite direction. Actually, it is better to drive both of the pump-gears from a single rotating source through a separate set of mechanical gears. If the left-hand pump-gear is turning clockwise, it will force water upward along the inner left side of the pump case and toward the outlet. At the same time, the right-hand gear will be turning counterclockwise, driving water upward along the right-hand side of the pump case. By making the gears fit tightly against the case and against each other, such a system can pump not only water but also air.

If a rigid-wall, non-collapsing hard-line suction or drafting hose is connected to the inlet of a rotary-gear pump, and the other end is dropped into a pool of water, as shown, it can theoretically pull a vacuum in the hose which should be able to suck water vertically 34 feet up the hose. (Why only 34 feet will be explained shortly.) In practice, if a pump can raise water even 25 feet above a water surface, it is doing very well. A rotary-gear pump is said to be self-priming, but should never be more than 15 to 20 feet above the surface of the pool from which it is to pull water to fill a fire engine's tank. These rotary-gear pumps can produce relatively high pressure but at low volume.

If water being drafted by a rotary-gear pump contains dirt, leaves, or other foreign matter in it, the gears or pump walls may become clogged or scratched, and the pump can be ruined. Another difficulty is the pulsating water flow produced by the discrete small pulses of water that are pumped. Such pulsations can cause undesirable vibration of the plumbing system unless a trapped air chamber, or dome, is added to the outlet line as in the early pumpers (Chapter 1). This acts as a cushion to reduce the pulsating effect caused by the pump. Although shown with only three lobes on each rotor, rotary-gear pumps may be made with many more lobes. This reduces pulsation strength and results in a smoother stream of water.

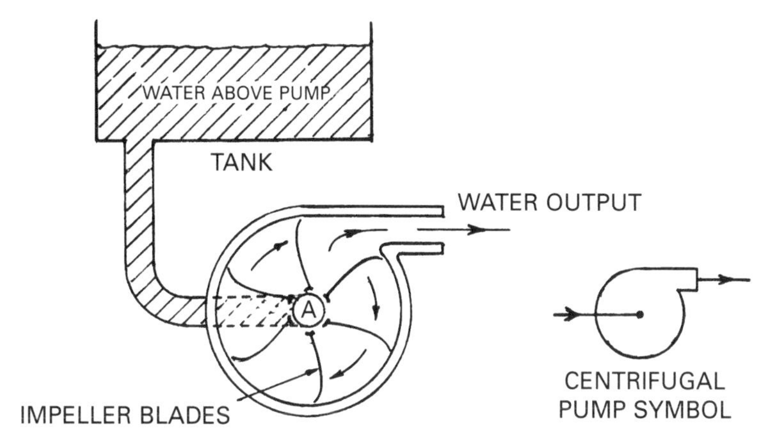

Figure 6-4. Basic centrifugal pump and symbol. Clockwise impeller blade rotation drives water around and out.

Centrifugal Pumps

High-volume fire engine pumps are centrifugal types, shown in simple form in Fig. 6-4. In this illustration, the circular rotating *impeller* mechanism has six blades and is driven at several thousand revolutions per minute. Any water introduced through the inlet at the center of the impeller blades (point A) is picked up by the blades and hurled outward from the center by centrifugal force. The outward ejection of water from the pump sucks more water into the center of the pump from the water source. The faster the impeller blades turn, the greater the volume of water discharged at the pump outlet, and the greater the water pressure that can be developed. This type of pump must be *primed* (filled with water) or it will probably not pump any water. If the water source is above the pump, as shown, gravity can provide the necessary priming to make such pumps pump water.

If the water source for a centrifugal pump is *below* its inlet opening, the pump must be primed by some external means. Note that from point A to the outlet, there are six spaces or openings, which means that there is no means of producing an airtight pump-starting action. Rotation of the impeller produces almost no vacuum on the inlet line and is not

considered to be self-priming. It can be expected that all centrifugal pumps will be installed in fire engines below the bottom of the water tank on the vehicle in order to provide a gravity priming of the pump by the tank water. Once primed properly, centrifugal pumps can draft water up to ten feet or more above the surface of a lake or other water source.

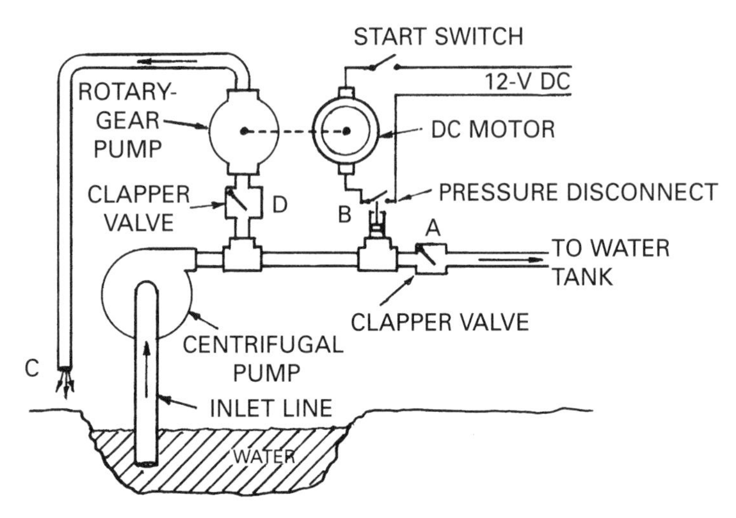

Figure 6-5. Rotary-gear pump circuit to prime a centrifugal pump.

When a fire engine's water tank is empty, it is necessary to prime the centrifugal pump in some way if water is to be drafted into its tank. This can be accomplished by having a small rotary-gear pump connected as shown in Fig. 6-5. Both pumps are started at the same time. The centrifugal pump is driven by the gasoline or diesel engine of the vehicle. The rotary-gear pump shown is being driven by an electric motor operating from the engine's battery. A clapper valve, shown at A, opens when water presses against it on one side and closes when water pushes against it on the other side. The valve at A is hanging down and therefore closes when the rotary pump starts to reduce the air pressure in the centrifugal pump and the pipes connected to it. This prevents air from leaking into the system through the pipe line from the tank.

As a vacuum is drawn on the inlet line through the centrifugal pump, water begins to flow up into the pump, through the rotary-gear pump and out onto the ground, at C. However, as soon as water fills the centrifugal pump, it is primed and begins pumping a large volume of water. The water pressure developed in the pipe line activates the pressure-disconnect switch, B, turning off the dc electric priming motor. The water pressure now in the system drives the clapper valve, D, closed. (The pull of any air drawn up by the low volume rotary-gear pump would be unable to close this valve.) The drafted water flows into the tank to fill it, or to fill all of the lines to the hose outlets, or both.

Because pumpers must be capable of drafting from pools of water, the fact that dirty water will not clog centrifugal pumps, plus the great volume of water they can move, makes them the desirable main pump for fire engines.

Two-Piston Pumps

The two-piston pumps explained in Chapter 1 were the pumps used in hand-pumpers and later in steam fire engines. If sand or other foreign substances were drafted with such pumps the sides of the cylinders become scored, ruining the pumps. While pumps of this type may not be desirable for high volume drafting, like rotary-gear pumps, they are suitable as a primer for a centrifugal pump, or for high-pressure pumping from the tank of a pumper to relatively small, ¾- to 1-inch diameter, high-pressure booster hoses.

Series and Parallel Pumping Systems

A single centrifugal pump can produce water pressures up to several hundred psi, depending on the number of hoses being fed water and the nozzle opening sizes being used.

Under a given load, a fire engine pump might produce 200 psi and a volume of 300 gpm. If two of these 200-psi 300-gpm pumps are operated on the same drive shaft, they may be connected either in *series*, for high pressure, to produce 400 psi at 300 gpm, or they may be connected in *parallel*, for high volume, to produce 200 psi at 600 gpm.

The diagram in Fig. 6-6a shows two pumps connected in a two-stage series system.

If each is capable of pumping 300 gpm at a given rotational speed with a given load, the two together can deliver only 300 gpm, but the pressure of one pump is added on top of the other, theoretically doubling the pressure to a restricting hose line and nozzle. In practice, the increase in pressure is closer to 70 percent than the hypothetical 100 percent more.

In Fig. 6-6b the two pumps are connected in a two-stage parallel system. The result is twice the volume of water flow to hoses, but little if any increase in pressure over what one pump would produce.

If a fire engine has two pumps capable of operating in either series or parallel, there will be a single control, called a *transfer valve* which, if turned, will change the plumbing to either a series or a parallel system.

If the engineer at the controls of a fire engine at a fire were to produce pressure and volume changes by switching a series-parallel system, it would be advisable to notify the firefighters, because a change of nozzles might be required for most efficient water streams. Higher pressure in hose lines could also make 1½-inch hand-line hoses more difficult for one person to handle.

Pumping Uphill

When fighting a fire several hundred feet up a hill above the only water source, one pumper parked near the source of water can draft water and pump it up to a second pumper stationed further up the hill. The second engine boosts the water pressure by using its pump to feed the water to either its hoses or to another pumper further up the hill. This is known as a *water relay*.

At high altitudes, pumps are less effective at drafting because atmospheric pressure, 15 psi at sea level, decreases about 1 psi for every 2,000 feet of altitude. At 10,000 feet, the atmospheric pressure on the water surface, which helps to push water up the drafting hose to the pump, is now only about 10 psi.

The weight of the air, or atmosphere, pressing down on everything at sea level is 15 psi. Why this is important and how it affects the drafting of water are interesting. Suppose we have a 40-foot or longer glass pipe with a 2-inch diameter, which is sealed at only one end. If it

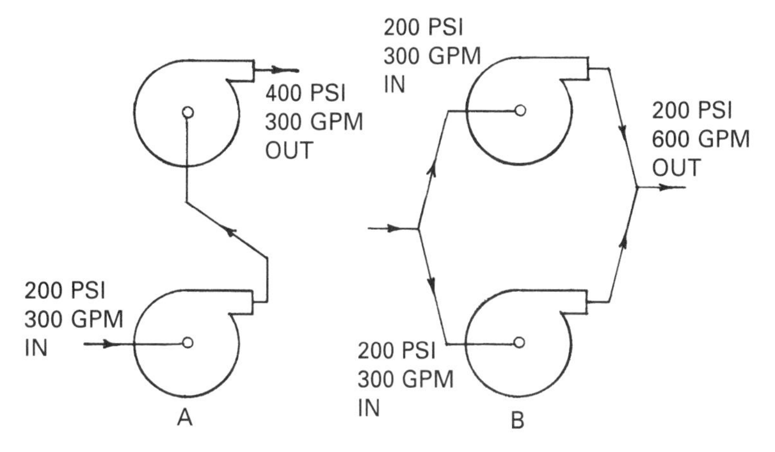

Figure 6-6. (a) Two centrifugal pumps in series for "pressure." (b) Two pumps in parallel for "volume."

is immersed in a lake near sea level, sealed-end down, it will fill with water. Now, if the pipe is rotated and pulled up out of the lake 30 feet (sealed-end up), the whole 30 feet of pulled-up pipe will be filled with water. If we continue raising the pipe until its top is 35 feet above the surface, we will find that there is 1 foot of empty space in the pipe above a 34 feet high column of water. As the pipe is pulled still higher, the space increases, but the water column stays at 34 feet above the lake top. The weight of the atmosphere on the lake top has pushed the water up 34 feet in the pipe. The space above the water column top is essentially an area of vacuum. This is true regardless of the angle of the pipe from the water surface—water would never be pulled more than 34 feet above the water surface at sea level. Do you think that increasing the diameter of the pipe would make a difference on the height of the water column? Not a bit!

When it is required to draft water 20 to 40 feet above a water's surface, as from a river up to a pumper or water tender that is on a high bridge, an *ejector*, Fig. 6-7, will have to be used. The ejector and its strainer are lowered over the bridge rail until the top of the strainer is a few inches below the water surface. The pump

drives a jet of water down and out through the constricted Venturi opening. As this water stream expands into the chamber ahead of it,

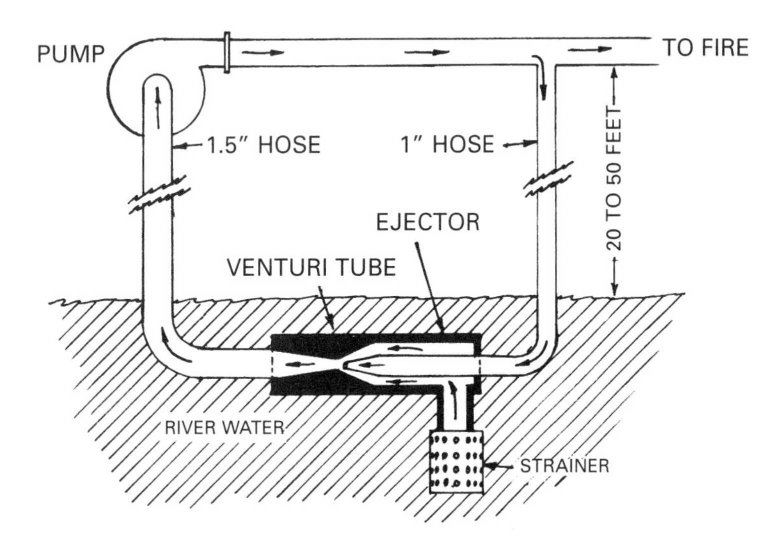

Figure 6-7. Ejector used when drafting about 20 feet or more.

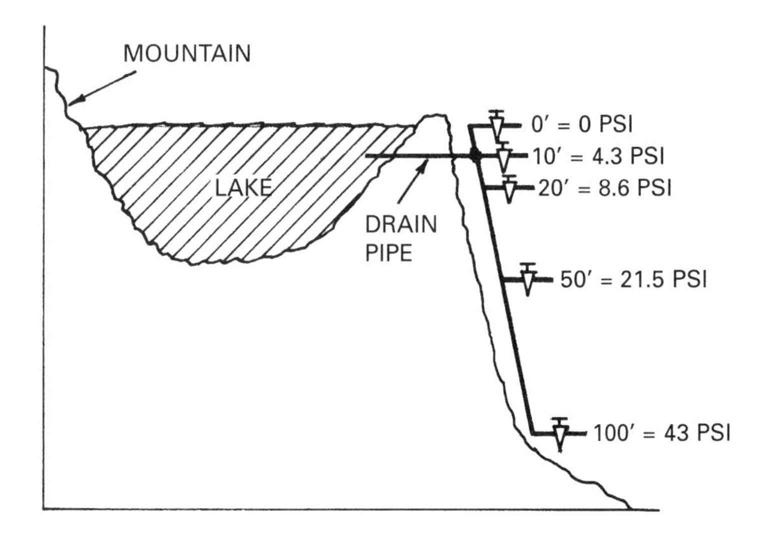

Figure 6-8. Various pressures developed with different water heads.

the partial vacuum, or area of reduced pressure, sucks water from the river into the ejector. Both pumper water and river water are now being delivered into the 1½-inch hose. Thus, water is both forced and drawn up to the pump where it is either pumped to the fire, or in most cases, to the tank of the water tender that is doing the drafting. Some water is lost to the output flow to keep the Venturi tube circuit operating at all times and to allow the water to be raised more than the 34 feet normal limit by a drafting pump.

Actually, water pumps are never 100 percent efficient and can rarely draft up to even 30 feet by themselves. They may require several minutes of pumping to raise water even up 20 feet.

While on the subject of the ups and downs of water, consider Fig. 6-8, a mountain lake with a drain pipe driven through the mountain edge to tap into the water 1 foot below the surface of the lake. Attached to this horizontal pipe is a 100-foot more-or-less vertical pipe, tapped with valved outlets every so often. Let's assume the top valve is opened and all others are closed. Now, the lake water can flow into the vertical pipe and fill it completely. There is only a dribble possible from the top valve outlet because it is at the lake's surface elevation, and there is essentially no water pressure here. However, for every foot below this, there is 0.43 psi of pressure developed. Thus, at the 10-foot level, at the second valve, there will be about 4.3 psi of water pressure available, and a low volume of water flow could be produced. At the 20-foot level, the pressure would be about 8.6 psi. At the 50-foot level, the water pressure would be about 21.5 psi. With 100 feet of *head* the pressure would be 43 psi, a low household water pressure value. The head would have to be about 233 feet to develop the common fire-fighting pressure of 100 psi.

The *static*, or no-water-flowing, pressure at 100 feet down the pipe might be 43 psi, but if the system is using small pipes, the actual dynamic, or water-flowing, pressure at the end of the small pipes would be less than that produced by 43 psi because of the internal resistance produced on the pipe inner surface plus any water turbulence developed in the small pipe.

You can apply the same reasoning to your own backyard. If you have a hose that extends 30 feet down a hill below a faucet that has a 50 psi static pressure, there will be a static water pressure at the closed nozzle of 50 + (0.43 x 30), or 50 + 12.9, or 62.9 psi. You will be able to

throw more water down there than you could at the faucet level. Would you get more gpm at the lower level than from the faucet alone? Probably, but that would depend on the amount of internal resistance in the hose and house pipes involved.

If a fire engine is pumping at 100 psi into an open 2½-inch hose laid up a hillside, how far up the hill could it deliver water as deduced from information above? Only about 232.6 feet, right? And how much water would it be delivering at that altitude? How about zero? At what elevation could it deliver 50 psi? About 116 feet, right? A relay pumper placed 230 feet above the first pumper would be able to relay water to two or three 1½-inch hoses or to another relay pumper 230 feet above it.

Slide-In Tanker

A handy piece of fire-fighting pumping equipment sometimes found on pickup trucks used in rural fire fighting is a *slide-in tanker*, Fig. 6-9. The gasoline engine, which ranges from two to ten horsepower, is directly coupled to either a self-priming type pump or to a small centrifugal pump that can be primed easily. (Some centrifugal pumps may be self-priming if the distance from the inlet pipe to the water surface is short enough.) The pump inlet is fitted with a 1½-inch pipe or hose that protrudes down to a quarter of an inch from the bottom of its 20- to 100-gallon water tank. A water pressure gauge is mounted on top of the pump. A reel of 150 to 300 feet of 1-foot booster hose is shown coupled to the pump through a rotary, or *swing joint* that allows water to be delivered to the hose while the reel is turning.

As soon as a slide-in tanker is parked at a fire scene, one of the firefighters can pull the charged hose out and move with it to the fire. The hose unrolls from the reel as he or she moves away. Another firefighter starts the gasoline motor and adjusts its speed to produce the desired water pressure (shown by the gauge). As soon as the first firefighter arrives within water-stream distance of the fire, the nozzle valve is opened and a straight stream, spray, or fog can be delivered at the fire within a few seconds. Note that there is a second gated outlet from the pump, allowing a second hose to be attached, or into which water may be fed to prime the pump. If pump priming is necessary, holding a few feet of charged line above the pump level with the nozzle valve open would allow any water already in the hose to flow backward into the pump and prime it.

We have discussed what might be carried on a fire engine and something about its pumps. Just how pumpers can do what they must do depends on their plumbing systems and the controls attached to them. The next chapter explains how a couple of possible, but relatively simple, fire engine systems work.

Question: How would the water pressure from your hose be affected if you hauled a hose up to a second-story porch to fight a fire there? Try to figure out the correct answer to this before reading any further. The answer follows, but don't look at it yet.

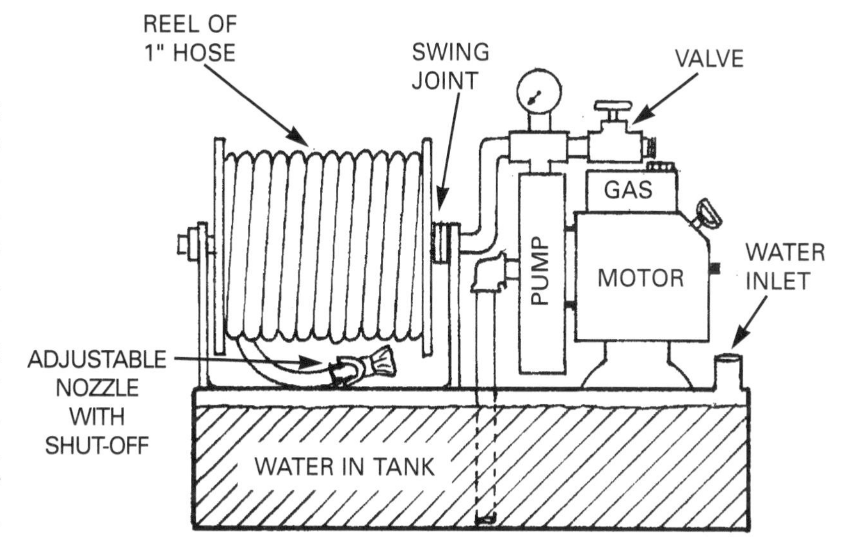

Figure 6-9. Essentials of a slide-in tanker with gasoline-driven motor driving drafting water pump.

Answer: The pressure would decrease 0.43 psi for each foot of altitude. A second-story floor is roughly ten feet above the ground floor. The rise in nozzle height over the first floor level and the approximate faucet outlet level would be about ten feet, plus four feet that the nozzle would be held above the second floor, or 0.43 x 14, or 6.02 psi. If you had 50 psi at the faucet, you would have only about 44 psi on a second story—and about 40 psi on a third floor.

7 Fire-Engine Pumping Systems

A Relatively Simple Pumper

Whether or not you will ever have an opportunity to actually operate a fire engine at a fire as a volunteer or paid firefighter, the functioning of the internal water system of a pumper is quite intriguing. It can be rather overwhelming when you first see the water-control panel of a fire engine outside and behind the driver's door. There is a mass of switches, levers, valves, capped hose outlets, gauges, and an engine speed control. Let's investigate what the basic plumbing system might be in a not-too-complex pumper.

Examine the schematic diagram in Fig. 7-1. In this case, the engine of the vehicle serves as both motive power for the wheels and drive for the centrifugal water pump. This may be either through a power take-off (PTO) from the transmission, as shown, or directly from the engine. In this particular case, it is possible to set both the vehicle's wheels and the PTO into simultaneous operation so that the pumper could move forward and also pump water on a roadside fire at the same time, often an advantage when fighting rural grass fires.

Depending on the available engine power of a pumper, usually in the 200 to 400 horsepower range, its water pump might have a capacity of more than 500 gpm at perhaps 200 psi. Such a pumper will be able to mount a reasonably effective attack on most home-type structure fires. In this simplified system it is shown with only two 2½-inch hose outlets. Adapters might be used to allow 1½-inch hoses to be coupled to these outlets if the pumper is located near the structure. If not, the necessary number of 50-foot lengths of 2½-inch hose would be stretched out along the ground to the fire. At the hose end a reducing *gated wye,* also known as a *dividing siamese,* would allow one or two 1½-inch lines to be coupled to the 2½-inch hose output to fight the fire from two directions. A second 2½-inch line with another wye would allow four 1½-inch hoses to be used on the fire. The more hoses used, the faster the engineer may have to run the vehicle's engine to maintain the necessary pressures to the hose nozzles.

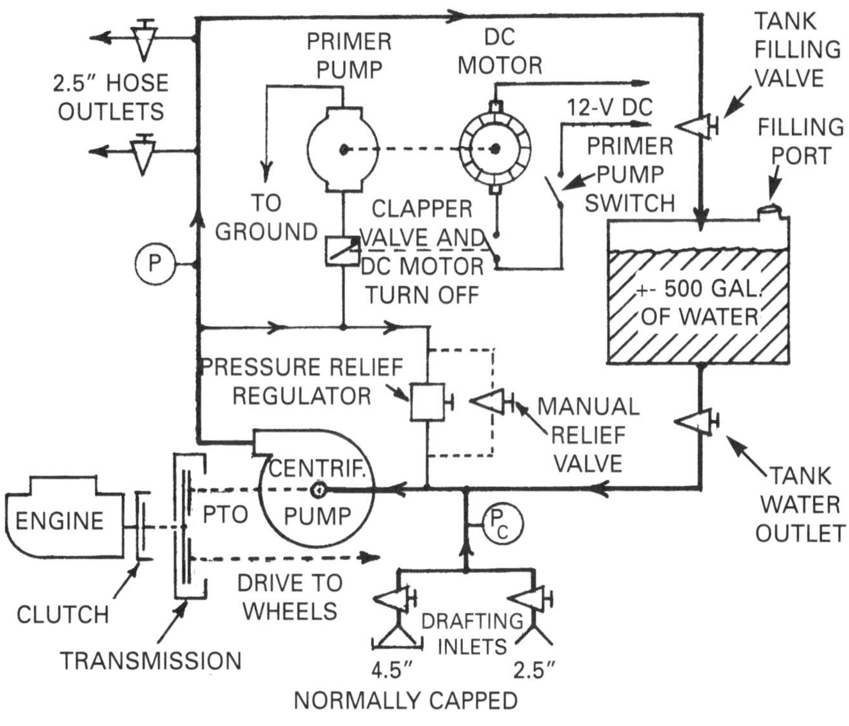

Figure 7-1. Simplified fire engine water system diagram.

Two possible types of pressure relief valves around the centrifugal pump are shown. One is a manual type, and the other is the normally used automatic type. With a manual type the pumper engineer must keep a close watch on the output water pressure and make an adjustment each time a hose is turned on or off.

Even when using an automatic valve, if the hoses are operating with relatively large-orifice nozzles on them they may bleed down the pump pressure so much that it may be necessary to increase the engine speed sometimes to increase the pump's output to maintain pressure to all hoses.

Possible methods of replacing water in engines that have used all of their water on a fire.

Two 1½-inch hand lines may use water at a 200 gpm rate (depending on the nozzle orifice) and a single 2½-inch line several times this fast. The ±500 gallons of water in the tank of a pumper can be depleted within a couple of minutes—perhaps five minutes if the water is used judiciously by the firefighters.

Replacing Water in the Pumper

One method of replacing water in the pumper's tank or to supply water to the fire ground is to run a hose from a hydrant, or water tender, to the 2½-inch female threaded *drafting* or *suction* inlet. (Many pumpers have a third [not shown], larger suction input pipe to connect to hydrants having larger output fittings.) With the hydrant connected and turned on, opening the *tank filling* valve passes the hydrant water through the operating or non-operating centrifugal pump. When the tank is filled, its filling valve is shut off. Some water can be delivered to the fire ground even if the pump is not operating, but much more, and at higher pressure, will be made available when the pump is turned on. Water can also be fed into the pumper's tank-top filling port with a garden hose or even by a bucket brigade (shades of the past!) to establish a prime if drafting from a pool of water is necessary.

Out in the country it may be necessary to draft water from a lake or swimming pool. The female threaded end of a 4½-inch hard-line will be connected to the 4½-inch male threaded inlet pipe on the pumper. The male end of the hard-line will be connected to the female threaded connector of a strainer. The strainer must be more than a foot below the surface of the water to prevent a whirlpool from developing, which would interfere with the suction action.

Providing Water with a Sump

A completely different method used to provide water for rural pumpers at long lasting fires is to use a *sump*. This may be a collapsible open-topped canvas or plasticized container with a capacity of several thousand gallons. Water tenders can rapidly dump their water supply into the sump while the attack pumper is drafting from it with a hard suction line. With such a sump, a few water tenders can provide a virtually constant water supply for a pumper even if water sources are miles from the fire scene.

Let's imagine a rural operation where a *nurse* water tender couples its discharge hose directly to a gated inlet of our pumper. The pumper's tank filling valve should be partially opened to allow filling of the pumper's tank at the same time that the pumper is delivering water to its fire-fighting hose lines. It is important to have the pumper's own tank filled again before the emptied water tender must uncouple from the pumper and go for a refill of its tank. In this way, a pumper will be able to continue feeding water to its hoses from its own tank until another nurse water tender's hose can be coupled to the pumper's valved inlet.

Sometimes water is used more slowly than might be expected because some of the firefighters may be shutting down their hoses as they knock down the fire at one spot and are moving over to another hotter spot. A properly made attack with fog nozzles on a lightly involved small structure fire may require a flow rate of 100 gpm applied to the fire for 30 seconds or less, or until the smoke changes to large volumes of white condensate. The larger nozzles may then be turned off, and smaller hand lines may be used to *overhaul*, or mop-up, what is left of the fire. At larger fires, the volume of water used will be in the hundreds of gpm and for long periods of time.

Drafting Water from a Pool

To draft water from a pool with our pumper, a hard-line would be coupled to the 4½-inch suction inlet. This inlet has male threads, so the female end of the hard-line would be screwed onto it. The male end of the hard-line would couple to the female threads of any strainer attached to the end that is dropped into the water source. The strainer would be floated at least two feet below the surface of the pool to prevent a whirlpool developing above it, which might draw air down into the strainer and interfere with the suction. Since a centrifugal pump cannot normally prime itself, all valves except the drafting inlet being used must be closed. The dc electric priming motor is then turned on. The primer pump pulls a vacuum on the closed pump system, causing water to rise via the hard-line into the main pump. It may take from a few seconds for a five-foot lift, to perhaps half a minute for a fifteen-foot lift. Once the centrifugal pump fills with water, it pressurizes the whole system. The establishing of pressure in the system automatically closes the clapper valve to the primer pump, which shuts off the primer motor. Water pressure builds up to the value at which the automatic pressure control is set by the engineer at the control panel.

Water from the pool can now be fed to the hose lines by opening the pumper's outlet valves. The tank filling valve can be cracked open slightly to refill the pumper's tank while the hose lines are in operation. There is always a slim possibility that the drafting line or the strainer might become plugged or stopped by leaves or other debris, and the pumper's tank water might then be needed until the drafting system is cleaned out.

Water Pressure

Pressure gauges are used to indicate the water pressure in the circuits to which they are connected. Gauge "P" indicates the pressure of the water feeding the hose lines. This will always be more than zero psi with the pump in operation. The "Pc" gauge, called a *compound pressure* gauge, has its zero calibration point somewhere above what would be the zero position on a normal gauge. This allows the pointer to move to the left of zero, indicating how much *negative pressure* (suction or vacuum) is being developed in the drafting system ahead of the pump. Atmospheric pressure at sea level will support a 30-inch column of mercury (or a 34-foot column of water). For this reason, negative pressure can be measured in inches of mercury rather than the psi numbers shown on the gauge for the positive pressure.

If water is taken from a hydrant or water tender through a soft hose, it is important that no negative pressure occur because such suction would collapse the hose and stop the drafting of water. In such a case, either the drafting pump speed must be reduced, or a hard-line must be used between the pumper and the hydrant or water tender that is supplying the water.

An excessive value of negative pressure at the inlet to the main pump may cause water-vapor bubbles to form on the pump impeller blades. These bubbles, thrown out to the tips of the blades, collapse as they move into a lower pressure area, producing shock waves and undesirable mechanical vibrations, termed *cavitation*. Cavitation can damage the impeller blades of a pump.

Pumper Circuits

A water system schematic diagram of a more representative pumper than the simpler one considered first is shown in Fig. 7-2. Many of the parts of this system and their functions are similar to those in the simpler pumper. The system shown here is not meant to be the same as any particular manufactured pumper but may resemble many in a variety of ways. Each of the many pumper manufacturers has its own plumbing circuitry.

This more advanced water pumping system consists of four basic sections:

1. *The water tank*
2. *A small-hose, high-pressure ("HP") system (±500 psi)*
3. *A large-hose, series/parallel, low-pressure ("LP") system (100 to 300 psi)*
4. *The truck engine*

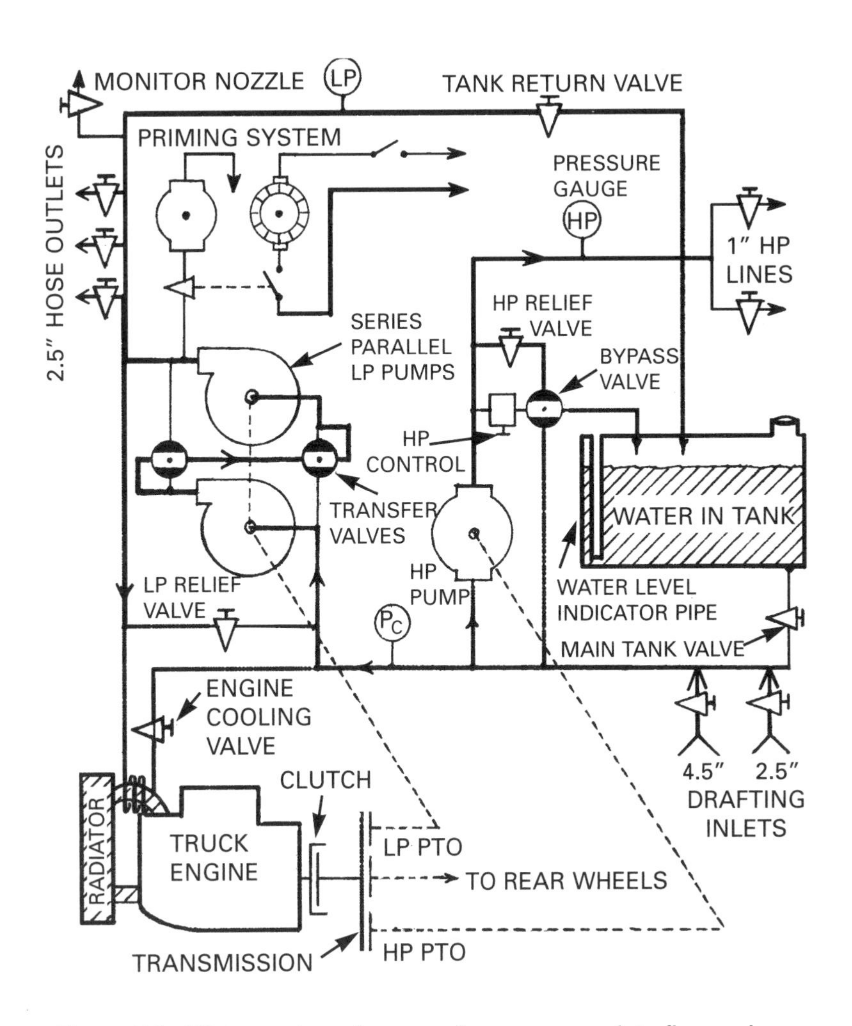

Figure 7-2. Water system diagram of a more complete fire engine.

Several basic uses and operations of the total system are rather interesting although complicated. Let's assume that water is coupled to the pumper but all valves are in a closed condition before the start of any of the operations listed below. See if you can follow the reasoning of each of the explanations of how water can be made available to pumper hoses to fight fires.

To pump tank water to large-hose LP lines:

The main tank valve is opened, and the centrifugal pumps are connected either in series for maximum pressure, or in parallel for maximum volume, by shifting the transfer valve setting. The truck engine is coupled through its transmission to the LP power take-off (PTO), which turns the LP pumps. The LP relief valve is adjusted to the desired pressure as read on the LP gauge. Hose lines can be coupled to any one or more of the three 2½-inch hose outlets. Water may also be fed to a rotating *monitor pipe* with a large-opening nozzle mounted on the top of the pumper. With the monitor nozzle, using the pumps in series, it is possible to throw an effective stream more than 100 feet at a volume of perhaps 750 gpm. Such a monitor would not be used if the only available water were that in the pumper's tank. It would only be used if the pumper were coupled to a hydrant. Opening the hose valves feeds LP water to the hose lines.

To pump tank water to HP one-inch hard-lines:

The main tank valve is opened, and the HP PTO is engaged. With the bypass valve in the position shown, as soon as the pressure at the hose outlets rises to the desired value (200 to 600+ psi), the HP control valve opens and overflow water passes through it back into the water tank. If the bypass valve is rotated 90°, the over-pressure flow returns to the HP pump inlet. Opening the HP hose valves feeds water to the hard (red or booster) lines and pressurizes them. The actual quantity and flow of HP water is considerably less than that of the larger LP hose lines, and it is controlled by adjusting the HP hose's nozzle.

To feed hydrant water to both LP and HP hoses:

The main tank valve remains closed. The bypass valve is rotated 90° to prevent hydrant water from flowing into the water tank. A 4½-inch or 2½-inch hard or soft line is coupled from the hydrant outlet to the drafting inlet of the pumper. The transfer valve is set for series or parallel LP pump operation as desired. The hydrant's valve is opened, and the LP and HP pumps are engaged through their PTOs. Opening the LP or HP line valves feeds water to their hoses. The pumps should not be run fast enough to produce a negative value as shown on the P_C gauge if soft hose is being used between pumper and hydrant or collapse of the hose may occur.

To fill the pumper tank with hydrant water:

A 2½-inch to 4½-inch hard or soft hose would be used between the hydrant and the drafting inlet. The main tank valve is opened, then the drafting inlet valve, and finally the hydrant valve is opened. The water level indicator shows when the tank is full enough to turn off the drafting inlet valve, the main tank valve, and if it was only required to fill the tank, the hydrant valve.

To fill the pumper tank from a pool of water:

Couple a 2½-inch to 4½-inch hard-line to a drafting inlet, dropping the other end of the hose with a strainer attached to it two feet below the surface of the pool. Open the valve of the drafting inlet being used. Set the transfer valves to either series or parallel, and engage the LP pump's PTO. Start the primer pump motor. When the LP pumps are primed, the primer motor should turn off automatically. A prime is usually recognizable by the change in sound of the pumps. At this time, the tank return valve should be opened and the tank should begin to fill. (It must not be open while priming because it will leak air into the primer inlet.) If the pumper is not going to be used to start fighting a fire, as soon as the tank indicator shows full, close the drafting inlet valve, then the tank return valve, and then turn off the pumps.

To feed water from a pool to LP hoses:

Couple a 2½-inch to 4½-inch hard-line to one of the drafting inlets, dropping the other end with a strainer on it two or more feet below the surface of the pool, if possible. Open the drafting inlet valve, engage the LP PTO, and start the primer motor. When the pumps are primed, the hose line valves can be opened to feed water to any hoses coupled to the outlets. If the tank is not full, the tank return valve should be cracked a little until the tank's water level indicator shows full, at which time this valve should be closed.

To feed water from a pool to HP hoses:

The HP pump shown is a rotary gear type and is self-priming. Couple a 2½-inch hard-line to the drafting inlet, dropping the other end with a strainer on it into the pool. Open the drafting inlet valve and engage the HP PTO. Open the HP relief valve to let pumped air and any water to escape to the tank through the bypass valve. When water begins to pump (the sound from the HP pump changes) the HP bypass valve should be rotated 90° to allow the pressure level to be set, and then the HP relief valve is closed. Opening the HP hose valves feeds water to the charged hard-lines. The HP hose valves may be left open even when the engine is not in use because charged lines must always be terminated with nozzles that have shut-off valves on them.

Radiator Cooling

When a parked, gasoline-engine pumper is operated at relatively high speed at the scene of a fire for an extended period of time, the engine's radiator may not be able to keep the

engine water cool enough. If the temperature rises above about 190° F, the engineer should open the engine cooling valve. This allows cold LP water to circulate from the LP pump outlet through a copper-tube coil device installed inside a piece of the radiator hose (shown outside for clarity in Fig. 7-2) between the engine and its radiator. The circulating cold water inside the copper tube accepts heat energy from the hot water returning from the engine to the radiator. The heated water is then led back into the LP pump inlet. This cools the hot water flowing to the radiator and lowers the gasoline engine's water coolant temperature. (The feeding of a little heated coolant water back into the pumper's fire fighting water supply is of no consequence.)

It can be seen that a system of this type can be quite complex, but it is also quite flexible. If the operator fully understands the operation of the system there will be other functions that can be used to advantage at a fire ground.

The explanations above have been more for equipment that might be used in rural fire fighting, mainly because it is simpler and a bit easier to understand. Fire engines in towns and cities may carry less water than those in rural areas because of the accessibility of water. It is expected that city engines will be operating within a relatively short distance of hydrants at all times. In general they have longer wheelbases, larger pumps, and carry more large-diameter hose. They are expected to pump more water over longer distances and to greater heights when a conflagration develops that involves all or most of a city block, or worse. It is also expected that they will be fighting fires in the upper stories of multi-story buildings. Regardless of these differences, the basic requirements of rural and much metropolitan fire-fighting theory and equipment are really quite similar.

It is easy just to say that a hose is "connected" to a pumper's valved outlet. But fire-fighting hoses and fittings are very different from those with which the general public is familiar. Just how they differ and what the general public may be able to do at a fire will be examined in our next chapter.

8 Fire Hoses and Fittings

Fire Hose Basics

Every home, particularly if it is out in the country, should have ready at all times at least a 75-foot long, 5⁄8-inch inside diameter, all-weather flexible plastic hose terminated with a combination solid-stream/spray/off garden-hose nozzle permanently screwed onto its *male* (externally threaded) end. It should be kept in a three-foot diameter roll in an easily accessible place. It should be left permanently attached to a centrally located outside hose faucet during hot, dry seasons. When not connected, the hose should be rolled up and stored somewhere handy. When storing the hose, push a cork loosely into the open *female* (internally threaded) end so that bugs cannot crawl inside, which would foul the nozzle when water is turned on. Shut down the nozzle for the same reason. If you live in an area subject to subfreezing temperatures, empty out any water in the hose before storing it.

The hose described above, with 50 psi from a home water system, will throw a solid 3⁄16- or 1⁄4-inch stream about 40 feet, and a 30° spray about 20 feet horizontally. From the ground, the solid stream will reach most of the roofs of small- or medium-sized single-story homes. If used while on the roof, or in a second story that is about ten feet above the ground, the pressure (assuming the nozzle is being held five feet above the floor level) will be reduced 0.43 x 15, or by about 7 psi, leaving a working pressure of 43 psi. Even if operated on a third floor, the water pressure will still be about 36 psi, which can deliver a reasonably usable spray for small fires.

While, as suggested above, a good emergency fire protection measure for a home is to keep one or more 75-foot, 5⁄8-inch garden hoses readily available, professional fire departments need a variety of more powerful hoses to adequately fight fires.

Some of the basics of fire hoses were discussed in Chapter 6. The smallest diameter practical fire engine hard-line type hose has a 3⁄4- or 1-inch ID and is made of either woven cotton, or some plastic material, such as nylon, rayon, or polyurethane, impregnated with rubber or synthetic rubber. These are used at working pressures from perhaps 200 to 800 psi, but are tested at three to five times the top expected working pressures. They will be 100 to 300 feet in length, and must be covered with a tough, abrasion resistant outer covering to resist damage when they are dragged over hot or rough surfaces at fire scenes.

There are several things you should keep available for your home's fire safety.

The next larger fire hose would be the 1-inch ID forestry hose. This hose is *single-jacketed,* lightweight, woven of mildew-proof linen or some synthetic and/or cotton cord. These hoses usually come in 100-foot lengths.

The largest fire hose normally expected to be handled by one person has a 1½-inch ID. It may be the highly flexible, single-jacketed, unlined woven linen fire hoses seen in glass-covered wall racks in large buildings. On fire apparatus, the 1½-inch ID hose is found in the form of a *double-jacketed* cotton or synthetic dacron, nylon, rayon, or polyester, rubber-lined hose. It comes in 50- or 100-foot lengths. With sufficient water pressure and large orifice nozzles, 1½-inch hoses can deliver as much as 150 gallons of water (three 50-gallon barrels!) per minute. Normally, they are used to deliver closer to 50 gpm. They may have a usable straight stream reach of at least 100 feet, and a 30° fog cone reach of perhaps 40 feet. The kind of nozzle used on it, and its adjustment, determines the characteristics of the stream of water delivered, its gpm, its reach, and its spray angle.

Stiff, steel-wire reinforced *suction* or *drafting hoses* are made in one- to six-inch ID sizes. They can be used to draft water uphill from a

body of water to a fire engine. There is also *soft suction* hose, which can only be used if the source of water, usually a street hydrant, has a greater pressure value than the suction value of the pump on the water tender or fire engine that needs water.

Many departments are using 1¾-inch ID hose, but squeeze it into the common, standard 1½-inch brass fittings. Today, because of cost, and to reduce weight, many so-called "brass" fittings are made of an aluminum alloy of some kind, such as Pyrolyte. By using this slightly larger hose, about 30 percent more water can be delivered, and the internal resistance to water flow is reduced. However, a larger diameter hose holds more water, is heavier, and may require two firefighters to handle it if the water pressure is greater than 100 psi.

How large of a hose can be managed by one person?

The standard 50-foot length of double-jacketed, rubber-lined 1½-inch hose weighs nearly 50 pounds empty and about 63 pounds when charged. (A charged 1¾-inch hose weighs about 73 pounds.) Three or four sections of 1½-inch or larger hose can represent quite a bit of weight for one firefighter to drag around. For this reason, hoses are laid to a fire while empty. As soon as they are in position, the firefighter at the nozzle might call back to the engineer, "Water!" and the valve to that line is opened at the pumper.

There is a variety of rubber-lined single or double-jacketed hose sizes larger than 1½-inch, such as 1¾-inch, 2-inch, 2½-inch, 3-inch, etc. The most common is the 2½-inch hose. There are also 3-inch ID hoses that are squeezed into standard 2½-inch brass or Pyrolite fittings. On long hose lays a 2½-inch hose will deliver about three times the water that a 1½-inch hose will, but it weighs more than twice as much when charged. One person trying to drag much 2½-inch hose over rough ground or up a stairway would find it either very difficult or impossible. And, according to "Murphy's Law," every metal coupling joint on the hose would be expected to hang up on every rock, stair, or projection over which it is dragged.

Nozzle Reaction

The *hose reaction*, or more correctly, the *nozzle reaction*, of large hoses makes it necessary to have one firefighter at the nozzle and another several feet back to raise the hose to prevent the nozzle from kicking upwards.

The subject of nozzle reaction is interesting. A nozzled hose, when fed high-pressure water, will attempt to straighten out. If laid unattended on a floor, it will whip back and forth dangerously. If you are holding a hose nozzle in which high-pressure water is flowing, you will find little trouble pointing the hose and nozzle upward, or to the right or left, but to depress the nozzle may require considerable energy. A 1½-inch hose and nozzle is manageable enough by one person, but a 2½-inch hose requires someone at the nozzle and someone else a few feet back to support the hose so the nozzle can be directed downward by the firefighter handling the nozzle. At 100 psi, for a depression angle of about 45°, a 1½-inch hose requires ±125 pounds of downward push to depress it, while a 2½-inch hose requires 375 pounds! Nozzle reaction of a garden hose is imperceptible under normal circumstances.

Hose-end Fittings

Standard fire hoses always have male-threaded fittings at one end and female-threaded fittings at the other end. Figure 8-1a shows a male-thread hose-end fitting at the top, and a female-thread hose-end fitting at the bottom. The upper part of the male fitting is a ridged tube. The ridging helps to lock the hose onto the fitting when an expansion ring is expanded inside the hose.

In general, *water flows out of male fittings.* On pumpers, all water outlet pipes have male fittings on them, with protective caps screwed over them while they are not in use. The female ends of *hoses* are connected to fire engine outlet pipes, putting the male threads at the far ends of the hose. Street water hydrants have male fittings under protective female screw-on caps.

In general, *water flows into female fittings.* An exception is the main water suction or drafting pipe on a pumper. It has male threads so that the female-threaded strainers that are attached to hard suction hoses used for drafting water into a vehicle will have female threads to screw onto the pumper's male threads.

Most fire hose *standpipes* mounted a few feet above street level on an exterior wall of tall buildings have female fittings on them so that hoses from pumpers with their far-end male fittings will couple to them. This enables water to be pumped up to the floors where water is needed. The plugs, or protective caps, for standpipes have male threads.

When there is a conflict with threaded fittings, there is a large variety of adapters, such as double-male 1½- or 2½-inch fittings, double-swivel 1½- or 2½-inch female fittings, to list a few. Figure 8-1b illustrates the difference between *pin lug* and *rocker lug* types of double-female couplers or connectors.

Normally, hose fittings should be twisted together only hand tight. If they can not be loosened by hand, then a *spanner wrench,* Fig. 8-1c, must be used on the pin or rocker lugs on the outside of fittings. Spanner wrenches are available in a variety of sizes and shapes, as well as the special spanner hydrant-type wrenches shown. Spanners are handy tools, sometimes being used as a light hammer, a window or door jimmy, or to turn off a gas or other type valve.

Snap type fittings are used by some fire departments. These fittings have no threads but are merely pushed together and lock when a detented half-turn is applied.

There are two basic wye (Y) types of hose fittings. One is called a *dividing type siamese.* It may have a large female input and two smaller male outputs, or all openings may be the same sizes, Fig. 8-2a. It can be used to divide the water from a 2½-inch hose into two 1½-inch hoses, for example. The outputs may or may not be shut-off valved (Fig. 8-2c). The other is a *uniting type siamese.* It usually has a large male threaded output and two smaller female input openings, although all openings may be the same size (Fig. 8-2b). It is used to accept 2½-inch hoses from two different pumpers and provide a 2½-inch or larger hose output to a fire ground, for example. Uniting siameses may have internal *clapper valves* that swing closed if one of the pumper lines weakens or stops. The greater pressure from the other pumper will push the clapper valve closed, and no water will go back up the lower pressure hose.

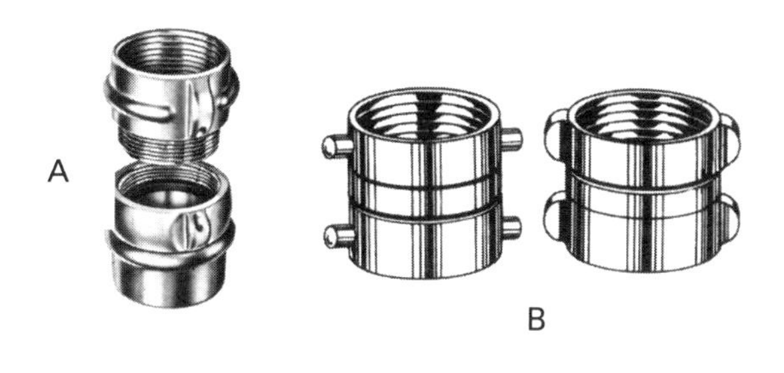

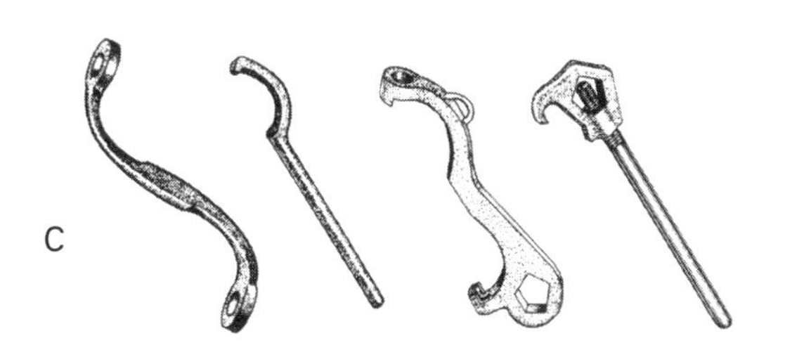

Figure 8-1. (a) Male hose fitting above, female below. (b) Pin and rocker lug double-female couplers. (c) Spanner wrenches used to loosen couplings.

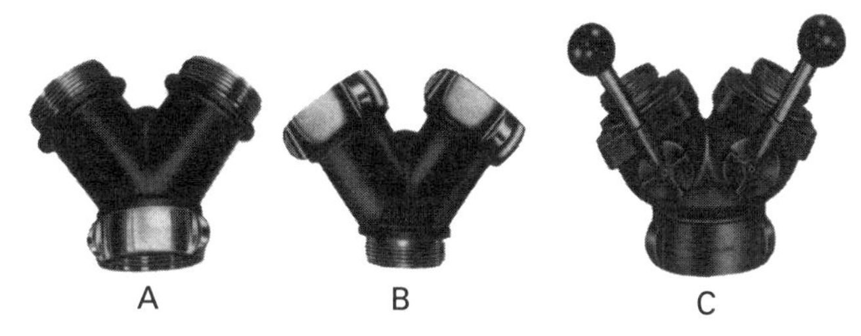

Figure 8-2. (a) A wye (Y) hose fitting. (b) A siamese fitting. (c) A double-gated wye fitting.

Another kind of a wye is called a *manifold,* a watertight box, or large pipe, having two to perhaps six gated outlets and one or more clappered inlets. A manifold can act as a centralized coupler for incoming and outgoing hoses at a fire scene.

Because of the many unforeseen events

that may require a variety of hose arrangements at a fire, each piece of fire apparatus has its own collection of double-male, double-female, reducing double-male, siamese fittings, plus manifolds, thread adapters, bushings, plugs, caps, and so on.

One would think that all fire departments would use the National Standard Thread (NST) fittings, but some may be using other types of threads, and some may use Jones or other snap-on fittings. Imagine the troubles that can occur when mutual aid fire equipment shows up from other nearby areas with different threads on their hose fittings. Thread adapters must be carried if such a situation is a possibility. Since most homes and buildings use *iron pipe thread* (IPT), some NST-to-IPT or IPT-to-NST adapters may also be carried by rural departments.

Some fire departments use sexless quick-connect *Storz* couplers which use identical fittings at both ends of their hoses. It is only necessary to push two fittings together, give them a 120° twist, and they lock together. Adapters for Storz-to-NST may then be needed.

Kamlok aircraft fittings, used on some air attack tankers, have two levers on the female fittings that lock onto the male fitting for rapid coupling and uncoupling.

As one might expect, eventually some hose sections will become burned, cut, or damaged to the point where they will not stand the required pressure test that must be applied to all hoses periodically. If a hose is damaged, but only near one of the end fittings, it may be possible to remove that metal end fitting, cut off the damaged part of the hose, and then replace the same fitting without shortening the hose more than a foot or so. However, this requires special tools and is not an easy task. It is done only if the hose is in good condition except for the one damaged end.

Now that we know something about hoses and some of their fittings, let's investigate some other important kinds of fire fighting devices that may be found at fire scenes in our next chapter.

9 Nozzles, Valves and Hydrants

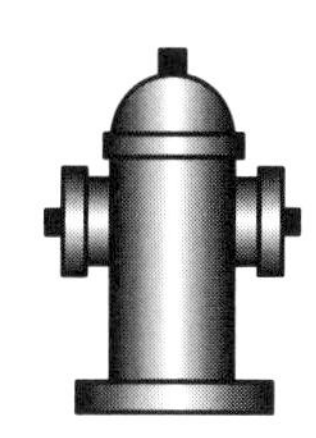

Nozzles

For more than 250 years, fire fighting nozzles were made of brass and constructed to deliver solid streams of water. They were called *underwriter playpipes,* because many of the original volunteer departments were fire protection or insurance companies, and the nozzles played a stream of water on the fires. The modern playpipe, Fig. 9-1a, may have two side handles and a shut-off valve. It is screwed onto the male threads of a 2½-inch or larger hose. Various sized tips may be available to screw into the front of the playpipe. If water pressure drops off, or if less water is needed, a tip with a smaller orifice can be screwed on.

Since World War II, fog nozzles, or, more likely, combination nozzles having four settings, straight-stream/spray/fog/off, have been in general use. Some may be a straight-stream/fog/off, with a control on the *fog* that produces a spray. A *straight-stream nozzle* is required when the fire is extensive or so hot that it is difficult to get the hose close enough to allow effective spray or fog operation. It is also used when necessary to drive water into narrow openings, or through windowed areas, or when fighting a fire in a lumber yard, etc. A large volume *monitor nozzle* on top of a pumper, or its ground mounted counterpart, the *deluge gun,* will usually use straight streams, although if they can be moved in close enough, a spray or fog type nozzle may be very effective.

When a very large volume of water is required, a deluge gun with a three-way siamesed inlet allows one, two, or three 2½-inch, or larger, lines to be coupled to it. If only one line is connected, the other two open female ports will be gated or have clappered valves inside them which swing down, closing off these ports when the water pressure inside exceeds the pressure outside. When other lines feed water to the clappered ports, the clapper valves are pushed open as a greater external water pressure is applied. With clappered ports, all of the multiple feeder lines must maintain approximately the same hose pressure in them.

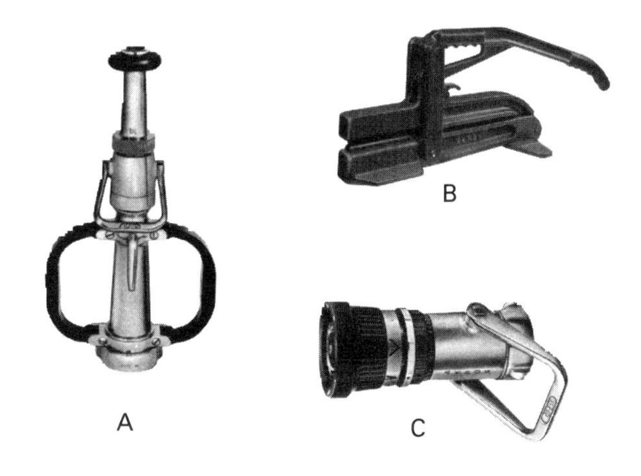

Figure 9-1. (a) Two-man playpipe. (b) Hose clamp. (c) Variable stream or combination nozzle.

Hand-held brass or Pyrolite hose nozzle openings vary in diameter from 3/16- to 3/8-inch for hoses having diameters of ¾- to 1-inch. Nozzles for 1½-inch hoses vary in orifice diameters from ½- to ¾-inch. Nozzles for 2½-inch hoses range in openings from ½- to 1¼-inches. Monitors and deluge guns may have orifices up to 2 inches.

To fight cellar, attic, or boat-dock fires, a nozzle with a special *multi-ported rotary head* can be used. It is pushed down or up through a hole drilled or chopped in a floor, ceiling, or deck. When water is fed to it, its rotary ports send out many small straight streams. Since the drilled outlets are canted slightly, water from them drives the head, which is on bearings, into rapid rotation, providing a circular, horizontal spray, as well as some spray both upward and downward. These *cellar nozzles* are not gated. To change from a fog nozzle to a cellar nozzle at

the end of a charged line, the fog nozzle is cracked open slightly to make it easier to unscrew the nozzle, and a metal *hose clamp*, Fig. 9-1b, squeezes the hose shut perhaps ten feet back from the nozzle. The fog nozzle is then unscrewed, and the cellar nozzle is screwed on in its place. As soon as the cellar nozzle is fed into the chopped hole, the clamp can be removed. If loosened before the nozzle is put through the hole, everyone in the vicinity may get sprayed—which may not be all that bad if the firefighters are close to flames.

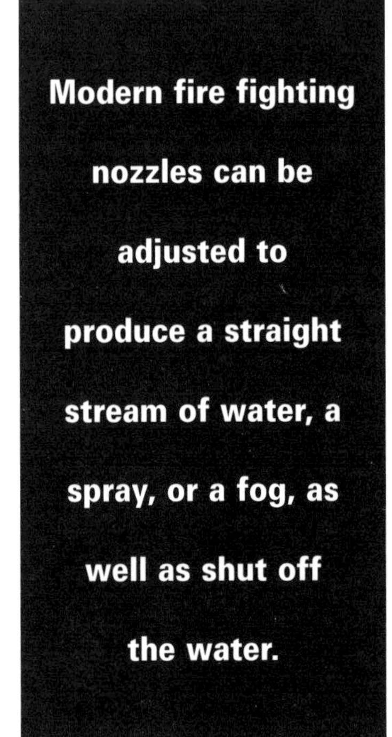

Modern hand-line nozzles do not break down into separate tip and valve sections as playpipes do. The valve handles, as in Fig. 9-1c, when pulled backward, open internal ball valves. If pushed forward, the internal ball valve shuts off the water. The handles are made *forward off* for a good reason. Sometimes a charged hose and nozzle is dropped at a fire line. As the hose and its nozzle are pulled backward from the fire, the nozzle shuts itself off as its handle rubs against anything on the ground or floor.

By rotating the forward section of modern nozzles, the water flow can be changed from a solid stream to a 30° to 45° spray cone, and then to a 90° fog cone. These settings are indicated by different width V's engraved on the barrel above the shut-off valve handle.

At 100 psi, a 1-inch hose will deliver about 12 gpm, straight-stream, spray, or fog, if it is fitted with a *constant volume* type of nozzle. A 1½-inch hose delivers about 22 gpm. A 2½-inch hose delivers about 35 gpm. This may not seem like a lot of water, but it is surprising how effective such nozzles are at fighting smaller structure fires.

Valves

There is a variety of valve devices used in the fire service. Two simple types are *in-line* (or simply *line*) valves, and *angle* valves. The valves in a gated nozzle, or in wye and siamese fittings, are examples of line-type valves. A simple line valve might consist of a short-threaded brass pipe-like body, with male threads on one end and female threads on the other, and some kind of a shut-off valve in the middle. It is not only used to couple two pieces of hose together, but also to cut off the water flow through the line when desired.

One fast-acting water shut-off device is the *ball valve.* It has a brass ball a little larger than the ID of the brass valve body in which it is located. The ball has a hole drilled through it the same size as the ID of the valve body. An external lever attached to the internal ball rotates the ball from a position of no opposition to water flow through it (hole in line with the flow), to complete shut-off (open hole at right angles to the flow). This is produced by a 90° rotation of the lever and ball. Care must be taken to shut down these fast acting valves slowly to prevent water hammer destruction of devices coupled to the water system.

When a round, flat metal disk is rotated inside an in-line valve body so that it can either impede the water completely, or not at all, by turning 90°, it is called a *butterfly valve.* A butterfly valve is also fast-acting.

A *gate valve* has a round handle which screws a flat, round metal disk down into a hollowed out channel directly across the flow of water in an in-line valve body. Gate valves are slow acting.

A *globe valve* has a round metal or plastic ball that can be screw-driven against the water opening in its pipe, stopping the flow of water. It is also slow acting.

Garden water faucets use *angle valves.* In them, a ball is unscrewed, backing away from its seated position across the faucet body. This allows water to flow over the ball into the internal faucet chamber and to the outlet set at right angles to the inlet. These are also slow acting.

A fitting that allows a smaller hose to couple to a main line to "steal" water while the main line is in operation is called a *water thief.* It is an in-line device with male and female threads at opposite ends, plus a reduced-diameter, gated (valved) male outlet on one side of

its body. The main line outlet (the male thread end) may also be gated. A forestry type of water thief system uses 1½-inch main line fittings, usually separated by 200-foot length hoses. It has a T-type one-inch side outlet that may be gated. To the one-inch outlets are coupled 100-foot single-jacketed *lateral hoses* with gated nozzles to put out, or to mop-up, the perimeter of grass, brush, or forest fires that have been knocked down by prior main line operations. The lateral outlet hoses may also be used to protect the main line hose itself if it becomes endangered by a later rekindling of the fire in its vicinity.

Hydrants

In cities, and in some rural areas, you may find one of two basic types of fire hydrants:

1. A *wet-barrel* hydrant, Fig. 9-2a, used when there is no possibility of water freezing in the vertical barrel part.

2. A *dry-barrel* hydrant, Fig. 9-2b, used where freezing of water in the barrel is a possibility.

In the wet-barrel hydrant, the smaller capped outlet might be for a 1½-inch hose, and the other outlet would be for a 2½-inch or larger hose. The barrel of such a hydrant must be at least 20 percent larger in cross-sectional area than the larger outlet. Each outlet of a hydrant is controlled by turning, with a standard hydrant wrench, the pentagonal *nut* end of the valve stem that comes through the opposite side of the barrel from the water outlet.

In the dry-barrel *frost-proof* hydrant, the valve stem extends from the below-ground main valve up through the top of the hydrant barrel. Screwing the below-ground valve downward into the water main closes off *weep*, or drain, holes as the barrel fills with water. When the valve is screwed upward to shut off water from the main to the hydrant, the weep holes are opened, and all of the water in the barrel drains off into the rock-filled area around the base of the hydrant barrel. Both the stem and barrel have break-away connections. If the hydrant is struck by a vehicle, these will break free at ground level, preventing a geyser of water from the broken hydrant. The hydrants mentioned here are just two basic types. Many different types are manufactured.

Many cities and towns may now be using the following recommended National Fire Protection Association (NFPA) standardized markings for fire hydrants. These are white reflective paint on the barrel of the hydrant for easy visibility at night, with a color-coded painting of the outlet caps of: *red*, less than 500 gallons per minute; *orange*, 500-1,000 gpm; *green*, 1,000-1,500 gpm; and *light blue*, 1,500+ gpm.

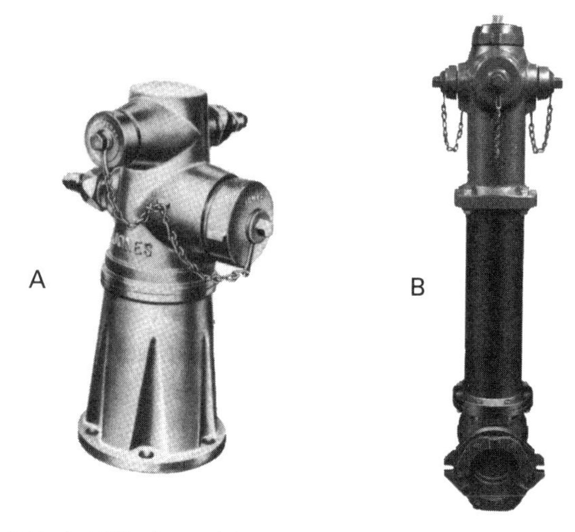

Figure 9-2. (a) Wet barrel fire hydrant. (b) Dry barrel fire hydrant.

Blue reflectors in the middle of streets indicate a nearby fire hydrant.

In cities you may notice small blue roadway reflectors mounted near the middle of the street opposite all hydrants. Firefighters seeing these reflectors ahead know they are approaching a hydrant.

The minimum size of the many modern water mains laid below ground to feed street hydrants is 6 inches ID. When there are several hydrants on docks or wharves, they may be single-valved, 2½-inch NST outlets coupled to 4-inch mains, although local regulations may differ.

Static water pressures in city hydrants may range from 40 to 100+ psi. However, if many hoses are playing water on a large fire, the operating hydrant pressures may drop consider-

ably below their static levels. This can be magnified by water resistance caused by roughness on the inner walls of underground mains, which may be miles long. By using a fire engine as a pumper, a city-main pressure of only 40 psi through a 6-inch main can easily be increased to more than 100 psi to feed several 1½- or 2½-inch fire hoses

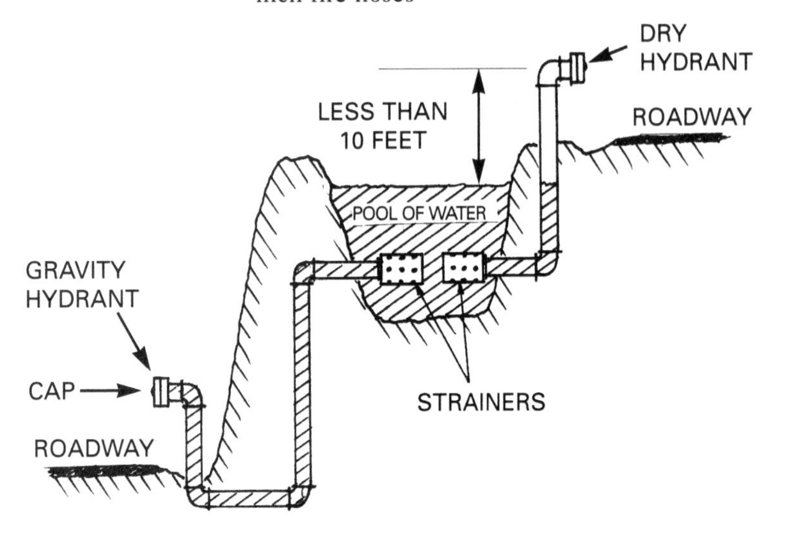

Figure 9-3. Possible types of rural gravity and dry hydrants.

In rural areas, some form of either a *gravity* or *dry* type hydrant may be found. The basic idea of these is shown in Fig. 9-3. The gravity hydrant pipe can feed water to a fire engine or tanker by gravity. It would be fitted with some type of shut-off valve. If the water head (the distance from the surface of the water source to the vehicle inlet) is not sufficient to develop adequate pressure, drafting with a pump in the vehicle will fill its tank more rapidly or pump more water to a fire. A strainer must be attached to the underwater end of all gravity hydrant systems. (Without it, the pump would deliver chopped fish, polliwogs, frogs, grass, etc. to the fire fighting hoses. None of these are particularly effective at putting out fires, not to mention their ability to damage pumps and to plug up hose lines and nozzles.) Strainers must be suspended above the bottom of ponds, lakes, or river bottoms to prevent picking up debris.

A rural dry hydrant may consist of an upstanding pipe, perhaps two feet high, terminated with a 1½-inch or larger NST male fitting at the outlet end. The far end of the pipe will be terminated with a strainer, usually in a lake. To get water from this kind of hydrant, it is necessary to couple a hard-line suction or drafting type hose to the top of the stand-pipe.

We have discussed some of the mechanical devices used by a fire department on its rolling equipment and at the scene of a fire. How about the firefighters themselves? How are they outfitted? What protection do they have? What must they wear? What equipment do they carry and use? If you were fighting a fire, what should you to wear to protect yourself? In our next chapter, let's discuss some protective gear used by firefighters.

10 Firefighter Clothing and Tools

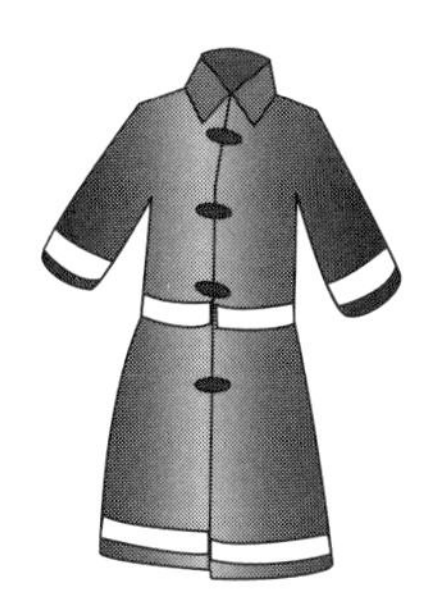

Turn-out Gear

The *bunker*, or *turn-out*, clothing worn by firefighters when responding to fires or other emergencies must protect them from water, heat, falling objects, nails and broken glass on floors, and in northern areas, from subfreezing weather.

Starting from the ground and working up, a firefighter should wear pull-on rubberized boots with puncture-proof non-skid soles, steel protected toes, and wool linings. The outside of all of the clothing worn should be resistant to common caustic chemicals. Most boots are worn without shoes, although there are some coverall boots that can be pulled on over street shoes.

Turn-out trousers should be loose fitting and are usually made of a synthetic material that must be heat-, chemical-, and abrasion-resistant. They must not stiffen in cold weather, and of course, must be waterproof. Although black does not show dirt as much as lighter clothing, and has been traditional, black surfaces may readily absorb radiant heat. Much of today's fire fighting clothing is bright yellow or aluminized and is fabricated of flame-resistant materials. Although lighter-colored clothing is harder to clean than black, it tends to reflect radiant energy as well as being more visible. A vapor barrier material may be inserted between the inner and outer surfaces of the material. An air space may be inserted between the inside lining and the vapor barrier layer.

The traditional bunker coats are made of the same black, yellow, or aluminized fabrics. Aluminized outer surfaces reflect radiant energy best, which keeps the firefighter cooler when near flames. However, when radiated energy from nearby hot fires strikes such clothing, it can still radiate through the material. The temperature build-up inside firefighter clothing can raise the person's body temperature from the normal 98.6° F to above 100° F, which can be dangerous to health. High reflective-index identifying stripes, Fig. 10-1, are usually sewn on all bunker coats to make the firefighter more visible at fires and other emergencies, especially at night.

Turnout coats have turn-up collars, several pockets, loops to hold tools, and four or five fast-acting clasps down the front. Coats extend down at least to the crotch and are loose

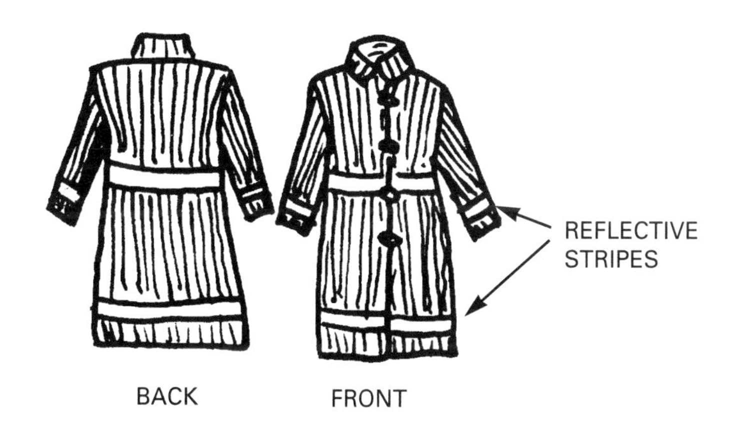

Figure 10-1. Basic turn-out coat with reflective stripes.

enough to allow free movement. For cold weather, inner linings may be used, and longer coats may be desirable.

Gloves are important safety items of clothing for anyone involved in fighting fire. Without gloves, handling hoses, nozzles, hot wood, fittings, etc., would cause innumerable scratches and burns. Firefighter gloves should have leather palms, preferably aluminized glass-fabric backs and should be long enough to protect the wrists when the hands are extended outward past the protection of the coat sleeves. When responding to a fire or other emergency, firefighters will either wear their gloves or carry them in a pocket. They may also carry small belt axes, a screwdriver, a hydrant wrench, and

a universal spanner tool. A pair of lineman's insulated gloves should be carried in each engine, and some firefighters should be trained in working with electricity.

The traditional fireman's headgear was the leather helmet with an identification of the fire company on its front, along with his rank, if the man happened to be an officer. Similarly shaped helmets are now available in light, tough, fireproof plastics or aluminum. The helmet is held away from the head by a headband slung on inner fabric strips to better protect the firefighter, should something hit the helmet. The brim is wide enough to drain water away from the face and the back of the neck. A clear plastic face shield, pivoted from the two sides of the helmet, can be lowered to protect the face, or be raised when not needed. Helmets are made in such a way that it is possible to wear an air breathing mask under them. A modern helmet is shown in Fig. 10-2. Special helmets can be fitted with an infrared sensitive thermal-imaging system that can perceive temperature variations as small as 0.5° F to reveal motionless victims, fire behind a wall, etc.

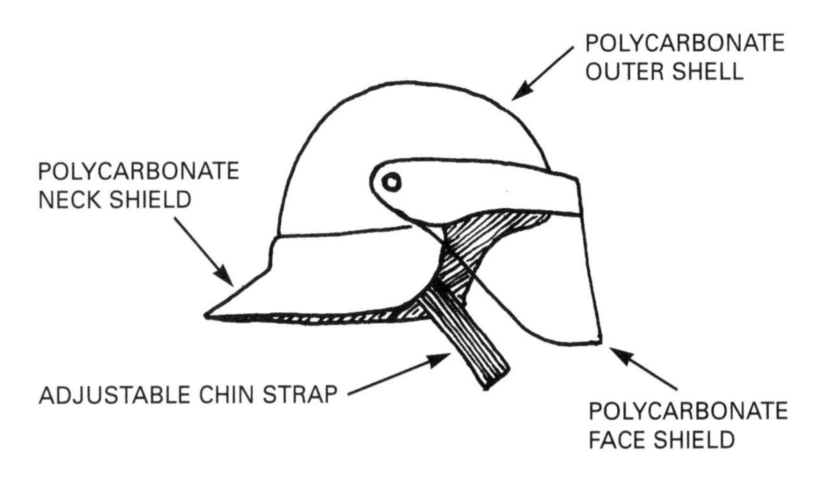

Figure 10-2. Fire helmet made of high-impact polycarbonate.

Breathing Apparatus

Inhalation of an atmosphere containing as little as 1 percent of the many poisonous gases which develop at most modern structure fires due to burning plastic interior furnishings can be deadly. For this reason, the day of the brawny, brave, smoke-eater type of fireman is gone. Without training and much experience, it is foolhardy for anyone to dash into a highly involved, smoke-filled modern-day building without portable fresh air equipment.

A firefighter's breathing apparatus is fairly simple and light. An airtight face mask is coupled by a flexible, heat resistant plastic tube to a tank of compressed air on the firefighter's back. Two looped straps on the tank assembly make it easy to sling the air tank onto the firefighters back in a couple of seconds. A belt passes around the body and is buckled at the front. The mask is put on the firefighter's face and under the helmet. The slight vacuum caused by a person inhaling inside the face mask opens the demand regulator valve on the air tank, forcing, at a reduced pressure, some of the tank's compressed air into the mask. As the person exhales, the regulator valve closes, and exhaled air is exhausted through a valved opening on the side of the mask. Although there are tanks with air supplies of seven, ten, fifteen, and fifty minutes, at most fires a more or less standard thirty-minute supply may usually be used.

It is general practice today for all firefighters going into burning buildings to wear a mask and compressed-air supply unit. The face mask improves visibility in smoke-filled rooms and results in far less eye irritation, and provides good physical protection for the face and eyes.

Tools

A *hose and ladder strap* is often carried to tie hoses to ladders. It is a strap with metal couplings at both ends. It can also be used, among other things, as a safety belt.

Pick-head fire axes are standard tools on most fire vehicles. They are used when forcible entry into buildings is required. They can be used to chop with the axe side of the head, and tear away walls, ceilings, floors or roofs with the pick side of the head.

One tool that has dozens of uses is the previously mentioned pike pole. It is a wooden

pole fitted with a steel head at one end. The head has a point at the top, with a hook just below, shaped like an old-time battle pike. The wooden handles may range from six- to eighteen-feet in length and are used for moving burning substances in both structure and wildland fires.

Used in the fires of early United States history, and still in use, is a *pull-down hook* (a large pike pole) with which roofing, siding, and weakened walls may be pulled down.

The *Pulaski* is a forestry tool having a two-sided head, one side an axe (vertical edge) and the other side an adze (horizontal edge). It is used for cutting trees and brush as well as trenching fire lines. An important forestry fire tool is a combination hoe and fire rake, called a *McLeod.* Another useful tool is a rubber or fabric beater, like a gigantic fly swatter, which is used to beat out small grass fires. It takes the place of the old-time wet sack, which, in a pinch, is still an excellent weapon against small grass fires, and even some small Class A surface fires. Unfortunately there are not many sacks available to the general public today.

When it is necessary to cut wire fences, electric wires, small bolts, etc., heavy-duty insulated-handled wire cutters are important firefighter's tools.

In some kinds of rescues, groups of power tools, such as the *Holmatro rescue tool*, or *Jaws of Life*, are used. They may be in the form of either hand-pumped (or other hydraulic-ram), or compressed-air operated tools. Smaller units can develop four tons of push, pull, lift, press, or spread, for victim removal from cars, or for moving heavy objects. Larger models develop over ten tons of pressure. Power-driven sheet metal shears are able to open overturned automobiles. They may be the only way entrapped victims can be quickly cut free and rescued.

Pulaskis and McCleods are two of the most useful fire tools used in fighting wildland fires.

Besides those mentioned above, there are many other common tools and equipment used by fire departments. Some of these are sledge hammers, crowbars, wrecking bars, shovels, brooms, tarpaulins of various sizes, ropes, rakes, hay hooks, cable cutters, buckets, and chain saws, as well as different forms and sizes of wrenches, pliers, and drills.

At fires or in any type of an emergency, one of the most important requirements is good communications—person-to-person—person-to-base-camp—base-camp-to-headquarters—department-to-department. Fire departments have kept up with these requirements very well. Let's investigate some of the ideas behind the complex modern-day fire communication systems in our next chapter.

11 Fire Alerting and Communications

Fire Alarms

If you happen to be the person who detects a fire, sounding the alarm might be as simple as a call of "Fire!" This may be relayed to a fire house by voice. Perhaps a nearby fire alarm box may be operated if in a city. It may be simpler to run to a nearby fire station or to call 911 by telephone.

Breaking the glass front on a fire alarm box in a building may either alert or allow alerting of a central point in the building, or the alarm may go directly to the local fire department, or both. Sometimes a fire alarm may be transmitted by a mobile or fixed Citizen Band (CB) radio station using the emergency Channel 9 (27.065 MHz) to a local sheriff or highway patrol unit.

In some cases, amateur radio operators may use their radio equipment for fire and highway emergency alerting and then may continue as back-up radio communications, particularly at wildland fires.

Actually, a telephone in a building, or a cellular transceiver in an automobile, or a hand-held cellular phone are the most generally used methods of communicating alarms. Telephoning a fire alarm would seem to be a simple enough procedure, and it normally is. In actual fact, however, lives and property have been lost just because of the improper use of a telephone. Too many times a telephone operator or a fire dispatcher has heard an excited citizen say, "Come quick! My house is on fire!" Click. If a rural operator does not recognize the voice, who knows where the fire is? With the 911 or other special telephone emergency systems, this should not be a problem. The address and owner's name is usually automatically displayed at the answering operator's position when a 911 call is received. But, the answering dispatcher will usually want to know: type of emergency (structure fire, grass fire, automobile fire, robbery, auto or other accident, etc.), the location of the incident, the nearest cross street, how to get there if in a rural area, is anyone hurt or endangered, a call-back number if the caller is leaving the telephone, and the name of the reporting person. Most coin-operated telephones allow you to dial 911 or the operator to report emergencies without depositing coins.

There are cases where telephoned fire alarms can be a tragedy.

A difficulty in telephoned alarms is having the receiving person jot down the correct street name or number. It is easy to transpose numbers or write down a wrong street name. For example, "6-8-2-5 Ellen Street" might be incorrectly transcribed as "6852 Helen Street." Always try to speak slowly, repeat all of your numbers, and spell the street name if it is possible that it may be misunderstood. In many cases people reporting an incident will speak with an accent. It may be difficult or impossible to understand them, particularly if they are excited.

When rural fires are reported, be sure to have someone at the entrance to the property.

If your home happens to be in a rural, hilly, or forested area, it is very important that someone be stationed at the driveway to the property where the incident is occurring to direct all emergency vehicles as they arrive at the driveway entrance. Too often the entrance is poorly identified, and mail box or street numbers are either hard to find, faded, unreadable, or missing. Be sure your street number is clearly visible day or night for any emergency vehicles looking for it. Even when smoke from a fire is in view from a rural roadway, it may be impossible for the responding firefighters to know how to get to the fire in the most direct manner. It helps, day or night, if you are waving a flashlight at an entranceway to signal emergency vehicles that you are the person to contact for routing instructions.

Some cities may still have street-corner alarm boxes. When these are opened and the internal lever is pulled down and released, a spring-wound motor may electrically tap out the location of the box, perhaps two and five. The next box down the street may tap out as two and six, and so on. This registers at the central alarm station and simultaneously at all fire houses in the city, even if the box is not in their area. (It is important that all nearby fire units be kept abreast of all fire alarms.) Reporting persons must understand that someone should remain at any pulled alarm box. They may have to direct the arriving engines to the fire, or whatever the emergency is, particularly if there is little or no exterior smoke showing from a fire.

As you travel the highways, you should note the different ways you could report fires.

The main disadvantage of street-corner alarm boxes is the number of false alarms that are pulled—as many as 50 percent in some cases. Because of this, many cities no longer use this kind of alarm, although it is an excellent system, particularly if the reporting person does not speak the language well, or at all.

In industrial or public buildings of all kinds, and in many homes today, sprinkler systems are installed in ceilings. When heat generated by a fire melts the heat-sensing fusible links, or *struts*, a valve is opened, and water is sprayed downward to put out the fire. At the same time, an electronic alarm may be set off or displayed at the security office of the building, or at the nearest fire station, or perhaps both.

There are also radio transmitter alarm boxes, often seen at the sides of highways. With one type, as the alarm lever is pulled down, a spring is wound. When the lever is released it springs back up, turning an internal generator that operates a small transistorized radio transmitter inside the box for a couple of seconds. The emitted digitally coded signal can contain information not only regarding the box location, but also if the emergency is a fire, or if police, ambulance, or automotive assistance is needed. This information is coded into the radio transmission when the reporting person presses one of several buttons on the box panel before pulling down the lever.

Some modern highway emergency call boxes are battery operated and may have a telephone in them connected by radio directly to some local emergency center. There will usually be a small solar-cell panel mounted a few feet above the box. The solar cells produce enough current when irradiated by the sun to keep the battery charged.

In some cases, it is advantageous to have several smaller fire departments operating in one geographical area, such as a county, using the same radio calling and working frequency. Adjacent fire districts will usually have *mutual aid pacts*. If a fire is reported in district A, it is probable that members in districts B, C, etc. will also hear the call. If it becomes apparent that the fire is getting out of control, the nearby departments may expect a mutual aid call. This is quite probable in rural areas where the only water at the fire may be that brought in by the engines and their water tenders. It is not unusual for several water tenders, and perhaps engines, to come in from adjacent fire districts on any sizable rural fire.

Radio Communication Equipment

Larger cities may be assigned their own fire frequencies, but their mobiles will probably have radio equipment that is also capable of working on all nearby rural district frequencies in case they are called in for mutual aid. There are often common state frequencies so that mutual aid engines from distant districts can communicate with those working in the local area. Some modern radio equipment can be adjusted to operate on any desired channel allotted to the fire service in a given *band of frequencies*.

Most of the radio communicating in the fire service uses *frequency modulation* (FM) to overcome the wide and rapid variation of received signal strengths when mobiles are in motion. FM receivers amplify received signals

The channels used for the Public Safety Radio Fire Service when this was written are in bands of frequencies from 33 to 459 MHz (Chapter 3). If you have a scanner receiver, you might try listening on some of these bands:

33.42 to 33.94 MHz	153.77 to 154.445 MHz
45.88 to 46.50 MHz	166.25 to 170.15 MHz
72.92 to 74.58 MHz	453.05 & 453.95 MHz
75.42 to 75.98 MHz	458.20 to 458.95 MHz

Some of these frequencies are for the use of *base stations* (fire department HQs) only, others are for mobiles only, and some are for both base and mobile use. In large cities the base station may transmit on one frequency, and mobiles answer on one or more other frequencies. Other bands will probably be allocated in the future.

to millions of times their actual received strength, then limit them to a more or less constant small fraction of their peak strength. Thus, FM signals can fade up and down over a wide range of strength, but the output voice signals are all limited to a given level and appear to have a constant loudness. This is true because the voice signals out of the microphones only *modulate*, or swing, the *frequency* of the transmitted carrier wave back and forth. They do not swing the signal strength up and down as broadcast band and television video *amplitude modulation* (AM) stations do. If AM signals fade very rapidly, the audio, or audible output, of the receiver can be very distorted. Furthermore, radio noise pulses, such as those generated by spark plugs in motor vehicles, are AM-type signals and interfere with reception of weak AM voice signals. The *limiter* circuits in FM receivers reduce such pulses so that they interfere little, if at all, with voice reception.

One effect produced in FM receivers that is an advantage (although it may be a disadvantage once in a while) is that the strongest FM signal tends to completely blank out weaker signals on the same frequency. If two fire engines, both on the same frequency, have the same signal strength at a receiver, both will be audible and possibly readable, like two people talking at the same time in the same room. If one of the FM signals is twice the signal strength of the other it will predominate, and the weaker signal may only be heard as a background noise. If one FM signal is perhaps four or more times stronger than another, it will completely blank out the weaker signal in the receiver. This can allow two different fire departments, operating perhaps fifteen miles apart, to operate on the same frequency with no interference to local signals. The distant department's signals may be heard whenever the local department's mobiles are not talking on the air. But as soon as the local mobiles come on, they blank out the weaker, more distant signals. In this case, every fire unit should identify itself each time it transmits.

Mobile FM receivers have *squelch circuits* in them. Only when a radio carrier (the part of a radio transmitter's signal that carries the voice modulation) is being received will the receiver's audio system open up to let voice signals come through. Squelch circuits take a fraction of a second to go into and out of operation. For this reason operators should not start talking until their push-to-talk switch has been down for perhaps a half-second. This is hard to do because when a person wants to say something it is natural to push the switch and start talking at the same time. But this may result in the first syllable or word not being transmitted and received.

Now being developed for safety service radio communications are digital transmitters and receivers. These scan, at perhaps 20,000 times per second, the relatively slowly varying (300 to 4,000 cycles per second) audio frequency alternating current (ac) that comes out of a microphone. The instantaneous voice voltage

value at intervals along the scan is converted into a binary digital number, such as 01001101. Another digital number is developed on the next instant, larger if the AC voltage is greater, smaller if the voltage is less, and so on. This is called *analog to digital (A/D) conversion.* The zeros and ones are transmitted by radio and picked up by receivers very accurately, therefore producing very little distortion. A circuit in the receiver converts the received numbers back into voltages equivalent to those originally sampled, reforming very accurately the audio frequency AC that came out of the microphone. There are a variety of advantages to transmitting and receiving digital signals, making them the communication method of choice for practically all communications in the future.

Using FM radio for fire fighting has some interesting advantages.

Mobiles should identify themselves in some way on each transmission they make using their *call sign* or *identifier (ID).* The spoken ID may be the mobile's assigned number, such as "1823," "Unit 1823," etc., or its department and type of vehicle, such as "Freestone engine one." This may take a little longer per transmission but can reduce confusion at times when a common frequency is used.

All of the bands used in the fire service are considered to be basically *line-of-sight* communication frequencies—if you can see the other station's antenna from your antenna, you should be able to communicate. But signals at these frequencies can be *reflected* from buildings, metal posts, wires, etc. Because of this, mobiles can travel through all sections of a city, even along streets lined by high-rise buildings, and although out of sight, may never be out of radio communication with other mobiles or the base station.

To some extent, FM and all radio signals may be *refracted* (bent) down over hilltops, or be picked up and re-radiated by wires, metal poles, etc. They may be received at the bottom of an adjacent valley, well out of visual sight of the transmitting station.

The lowest of the fire FM frequencies are sometimes affected by *long skip* conditions. (You may also notice this on the lower numbered TV channels if you are not using cable.) At these times, distant transmissions, up to many hundreds of miles away, may wipe out a weaker local station! This is caused by the refractive or reflective action of the Heaviside layers of the *ionosphere* (thin gaseous layers above our breathing atmosphere). Radio waves traveling through these layers will be bent back toward earth. At certain times of the year, the layers form in such a way that many frequencies below perhaps 100 MHz return to the earth's surface several hundred or maybe thousands of miles away. Frequencies above 100 MHz will only rarely be bothered with long skip interference. Once in a while, relatively low altitude hot or cold layers in the lower atmosphere form *ducts*, or special layers, that can funnel radio signals a few hundred miles before they can return to earth. These, too, can interfere with some local signals.

When mobiles are not out in clear areas they may be affected by *multi-path signals.* This occurs when the direct signal from a base transmitting antenna strikes the mobile antenna, but at the same time, the same signal bouncing off of some nearby reflecting surface is also picked up by the mobile antenna. Such multipath signals may be "in phase" and be additive in strength, or they may be "out of phase" and tend to cancel each other. As the mobile moves, the rapid phase shifts result in a rapidly fading, fluttering, or *picket fencing*, of weak received signals. If a mobile unit is parked where the signal from another station seems to be unusually weak, moving the mobile only one or two feet forward or backward may result in receiving a much stronger (in-phase) signal. There is nothing that can be done about these occurrences. Luckily, all equipment at one fire ground will be able to produce enough signal in all the local receivers so that skip and multi-path interference will not be bothersome.

Only vertical whip antennas are used for mobile communication in the fire service. This kind of antenna radiates and receives equally

well in all directions. A horizontal antenna will transmit and receive in some directions far better than in others. If horizontal antennas were used, by turning a corner the signal to or from a fire engine might fade out completely. Your fire-frequency radio scanner should always use a vertical antenna.

Repeater Stations

Because of the line-of-sight characteristics of fire radio frequencies, the higher and the more in the clear a transmitting and receiving antenna is, the greater the coverage area that can be obtained with good signal strength. Since it is unlikely that a fire department's main operating position for its base station will be on the highest point of land in a district, the base transmitter, receiver, and antenna (now operating as a *repeater station*) can be installed on a nearby mountain, hill, or perhaps on top of the highest building in town. Microphone audio (voice) signals can be fed up to the transmitter, and the received audio signals can also be fed down to the fire station control point by direct private wires, or by leased telephone lines.

When a department uses a radio repeater station, communicating distance may be increased, perhaps by more than five times. But now there may be interference caused by a second department's mobiles operating on the same frequency, although relatively farther away geographically. They may put in a stronger signal to the repeater receiver than the local department's own mobiles. One of the departments may have to shift to some other working frequency (channel) if interference is excessive.

When a repeater is used, the operator at the control point pushes a switch and talks into the local microphone. The voice signals are sent up to the transmitter on the hill by wire lines and are transmitted to mobiles anywhere that is radio-visible from the peak. When the switch is released, answering signals picked up by the receiver on the hill are fed down by wires to the operator's position.

Another common method of sending voice signals up to and down from a remotely located repeater is by using a microwave relay system. Microwave is a radio system using very high radio frequencies and directional *parabolic* or *dish-shaped* antennas. When a dispatcher at the fire station speaks into the microphone, the voice signals will modulate the microwave transmitter on the roof of the fire station. These signals are beamed up to the remote receiver dish on top of the hill, where they are made to modulate the repeater's fire frequency transmitter to radiate signals to mobiles. Received signals from mobiles are made to modulate the microwave transmitter up at the repeater site, which are beamed down to the microwave receiver on the roof of the fire station, are demodulated (made audible), and are fed to the dispatcher's loudspeaker or headset.

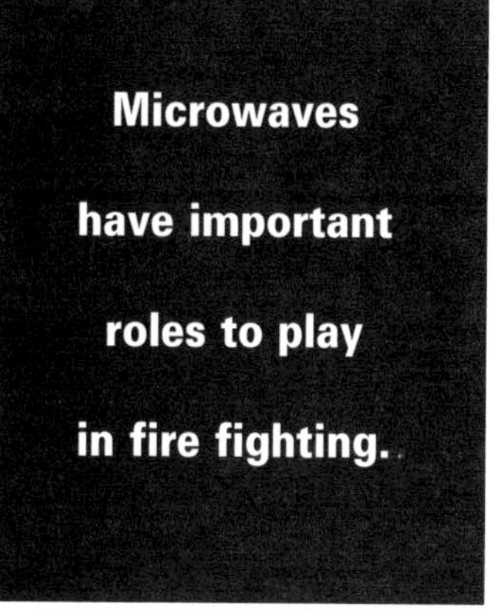

Remote microwave radio systems are particularly desirable for large city, county-wide, or state-wide fire communication systems. For a smaller town, a remote transmit-receive system might not be desirable at all. An FM transmitter and receiver on top of the fire house of a small town would be less likely to be blanketed by distant-city mobiles and would be far less expensive to install and maintain.

All modern mobile fire radio equipment, transmitters, and receivers, are transistorized. Mobile transmitters may radiate power levels of 10 to 100 watts. Base stations and repeaters may radiate up to about 250 watts.

Microphone Operations

If properly adjusted, a mobile microphone requires that the operator speak in a reasonably loud voice, with lips no more than an inch or so from the microphone. If the mobile operator speaks softly, or holds the microphone too far from his or her mouth, the signals being received at the other end will not be loud enough, and all the local extraneous wind, traffic, siren, and other noises may drown out the weak-modulation voice signals completely.

At the base station, the microphone gain

should be set so that adequate modulation is produced only when the operator is within one or two inches of the microphone to make sure his or her voice overrides background noises such as sirens, bells, telephones, typewriters, and nearby conversations.

Fire engine radio receivers often have an external loudspeaker connected across their output through a switching circuit. When firefighters are out of the vehicle, the external loudspeaker is turned on, allowing any calls made on the fire frequency to be heard all over the fire ground.

Many hand-held transceivers are now used at fires by officers. With these, almost everyone can be in instantaneous touch with the chief or the incident command officer. Fire helmets have been developed with low-powered transceivers inside them to enable working firefighters to be in constant communication with the incident commander. Such equipment is usually voice activated (stays off the air until the person's voice is picked up by the lipedge microphone in the helmet).

Cellular telephones are often being used for communications from the fire ground to persons who are not normally involved with fire activities.

A portable two-to-fifteen watt voice amplifier (*loudhailer* or *bull horn*) with a microphone at one end, a horn loudspeaker at the other, and batteries in the handle, is a welcome addition to any officer's equipment at a fire. It can be used to guide firefighters at a distance, advise occupants in a burning building what they should do, direct traffic, and so on. These are far better than the original speaking trumpets, or megaphones, used by early fire department officers.

In the past, number codes were used by many departments, such as 10-4 meaning "OK" or "I acknowledge." Today most fire departments are going back to the use of plain language, called clear text, and using common phrases. Some examples of self-explanatory clear text you might hear on fire channels are: Unreadable; Loud and clear; Stop transmitting; Copied (OK, or Roger, may be preferred);

Considering what we have been reading, a "Ten Commandments" for fire radio communications might be:

I.Thou shalt assure that thou hast received and transmitted all thy messages correctly.

II. Thou shalt speak into thy microphone distinctly, slowly, not shouting, from no more than two inches.

III. Thou shalt always identify thy mobile unit properly when communicating, and when at a base station thou shalt ID properly after each exchange of communications, or once each thirty minutes, as preferred by thine own fire chief.

IV. Thou shalt always call the other station first and then say, "this is" and then giveth thine own ID.

V.Thou shalt make sure that all other stations are off of thine own frequency before thou presseth thine own push-to-talk microphone switch.

VI. Thou shalt wait one half-second after pushing thine own push-to-talk button before starting to talk.

VII. Thou shalt advise other stations on thine own frequency who art handling non-emergency traffic to stand by until thine own emergency traffic is completed.

VIII. Thou shalt use the prescribed operating procedure, wording of information, or coding, as set forth by thine own fire department.

IX. Thou shalt not make tests of any kind on any channel on which voice traffic is being handled.

X.Thou shalt not transmit obscene language, superfluous signals, false or deceptive information, or interfere with any distress or emergency traffic—so sayeth ye olde Federal Communication Commission!

Affirmative; Negative; In-service; Out-of-service; Repeat; What is your location?; Identify; Stand-by; At scene; Available at scene; Can handle; Report on conditions; Fire under control; Returning to quarters (or station); In quarters; Resume normal traffic.

Phonetic Alphabet

To confirm the spelling of unusual words that are transmitted, it is recommended that some accepted standard *phonetic alphabet* be used. One that may be employed is the International Phonetic Alphabet:

A = Alpha	N = November
B = Bravo	O = Oscar
C = Charlie	P = Papa
D = Delta	Q = Quebec
E = Echo	R = Romeo
F = Foxtrot	S = Sierra
G = Golf	T = Tango
H = Hotel	U = Uniform
I = India	V = Victor
J = Juliet	W = Whiskey
K = Kilo	X = X-ray
L = Lima	Y = Yankee
M = Mike	Z = Zulu

Using the international phonetic alphabet, the word "fire" would be spelled, "**F**oxtrot-**I**ndia-**R**omeo-**E**cho."

Many areas may get together and develop their own standard phonetic alphabet. An example of another phonetic alphabet in use is:

A = Adam	N = Nora
B = Boy	O = Ocean
C = Charles	P = Paul
D = David	Q = Queen
E = Edward	R = Robert
F = Frank	S = Sam
G = George	T = Tom
H = Henry	U = Union
I = Ida	V = Victor
J = John	W = William
K = King	X = X-ray
L = Lincoln	Y = Yellow
M = Mary	Z = Zebra

To and From a Fire Scene

When dispatched, the driver of a fire engine should:

- *Carry his or her driver's license.*
- *Wear turn-out gear.*
- *Turn on the radio equipment as soon as the engine is started.*
- *Immediately notify the base station: "Central, this is XXXX In Service." (Many city departments use, "Out of service" to indicate that units are on the way and not to be assigned to any other location.)*
- *Turn on red lights and siren/yelper.*
- *Drive at speeds not exceeding the law, being careful when turning corners, and slowing drastically when driving through red-lighted crossings, stop signs, and when crossing railroad tracks.*

On arrival at the fire ground, each unit should notify the base, "Central, this is XXXX at the scene,"or perhaps, "Central, this is XXXX available at the scene." (Wording differs with different departments.) If an external loudspeaker is used on the vehicle, switch the receiver to feed its signals to the outside speaker.

As soon as the fire has been knocked down and mopping up begins, the Officer in Charge should have one of the units notify the base, "Central, this is XXXX. Fire under control." This is important to allow the frequency to be used once more for any non-emergency traffic. In many cases, vehicles at a fire ground will shift to a secondary channel for fire ground operations. When a unit starts back to its station, the base should be advised, "Central, this is XXXX returning to quarters." When arriving at the station house, the base can be notified, "Central, this is XXXX in quarters." When parked, all radio and electrical equipment must be turned off before the driver leaves the driver's seat.

When big city fire equipment is digitally dispatched, the destination of the alarm is printed out on a sheet of paper at the fire house. In each responding mobile unit, there will be a control box, perhaps with five buttons. As three of the buttons are pushed, they turn

the radio equipment on or off and send information by radio to the base station where it is printed out, along with the time. The controls might be:

1. *Turn on the transmitter/receiver (the transceiver).*
2. *Send the digital ID of the vehicle and "Out of service enroute to fire."*
3. *Send the digital ID and "Arrived at the fire scene."*
4. *Send the digital ID and "Vehicle enroute to the station."*
5. *Turn off the transceiver.*

Besides bells, sirens, or horns mounted on some centralized building being used to alert volunteer firefighters of a fire or other emergency, volunteers may be called by telephone or alerted by radio receivers at their homes. In many areas firefighters may be issued tiny pocket-held or belt-held receivers, beepers, or transceivers that must be on at all times.

At grass, brush, and forest fires, fire engines may not be able to hold fixed locations. As the fire progresses, personnel and apparatus may have to follow the fire. Use of radio communications in this case is most important. The coordination of operations of several crews out of sight of each other in a forest or in grassy wildlands can most easily be accomplished by using handheld radio transceivers, aided by officers possibly in helicopters or in fixed-wing aircraft above the fire scene, communicating with ground personnel by radio, as we shall see in our next chapter.

12 Wildland Fires

What Are Wildland Fires?

The term *wildland* in the fire service means a wild grassland, a brush-covered area, a forest, or a swampland. If you live in a rural area, or even on the edge of cities or towns, the information in this chapter could be of great importance to you. Wildland areas are usually not cultivated or inhabited to any great extent. Fires in such areas will normally be Class A types. Anyone who happens to be in a wildland area in which a fire is under way may be required to "volunteer" (?) to help fight the fire.

In our wildland areas a long-period ecological cycle is continually under way. The various phases are:

1. *A forest is burned to the ground.*
2. *A grass field grows up in its place.*
3. *Brush begins to grow in spots, shading some of the grass thereby killing it off.*
4. *Seedling trees begin to grow up through the brush.*
5. *Trees develop shade that kills off the grass and brush.*
6. *Small fires burn out dead underbrush and lower branches beneath the now fully grown trees—a forest is reestablished.*

This cycle may be repeated every 100 to 200 years or so, depending to a great extent on the latitude of the area.

Each one of the above six phases may require a different attack if a fire starts in these areas.

Fighting structure fires may be considered similar to old-time wars in which the enemy is relatively stationary. Structure firefighting apparatus rarely moves after the initial fire-fighting position has been established. On the other hand, wildland fires may travel rapidly and may require a running attack when they are extensive. While a water attack may be the best method of extinguishing Class A fires, adequate quantities of water may not always be available in wildland areas. The job of the firefighters at these fires may be to try to put out the edges or flanks of the fire, pinch it off, isolate it in some way, and let it burn itself out.

Should you ever be involved in a wildland fire, keep an eye on the wind, or you may not be around to fight another fire!

Structure fires are fought with water, but wildland fires in remote areas may often be fought with shovels or earth-moving equipment, *Pulaski* and *McLeod* raking/chopping tools, axes, saws, water or chemicals dropped from the air, back-fires, and limited amounts of hand-carried water. In structure fires, wind and weather are factors to be considered, but in wildland fires the wind and weather may well be the paramount factors. If you should ever be involved in fighting a wildland fire, you had better keep an eye on the direction taken by the wind, or you may not be around to fight another fire!

In the western United States, the rains usually drop off sometime in May. By mid-June the wildland growth can be bone dry, and the high fire danger season starts. The fire services keep a close watch on the relative humidity (percentage of water vapor in the air) and the areas of high and low barometric pressures. Air-mass movements (winds) usually flow from areas of high pressure to low pressure. In northern America, winds rotate clockwise around high-pressure areas and counterclockwise around low-pressure cells. Winds from the ocean may bring fogs, clouds, and higher humidity air. These allow fuels, grasses, and brush to retain their water, slow their burning, thus reducing the fire danger.

Winds from the mountains, inland valleys, or deserts may carry almost no moisture. With these winds, fuels dry rapidly, and the *fire danger index* rises. In southern California, for example, high-desert winds called *Satanás* (Satan in Spanish), *Santanas*, or *Santa Anas*,

blow from the mountains down toward the ocean, picking up heat as they travel downward. They may attain gusts close to 100 mph and develop temperatures well above 100° F. Wildland fires under these conditions are difficult to fight until the winds die down, or the fires burn themselves out.

In the East, Midwest, and South, the summer months bring many rain squalls, allowing wildland growth to hold its moisture until later in the year. The high fire danger season may develop during the late summer or early fall months and sometimes even later.

Causes of Wildland Fires

Although most forest fires are caused by lightning strikes, many of these are put out by the rain, if any, that accompanies the storms. There are many causes of wildfires. Some of these are:

- *Carelessly flipped cigarettes*
- *Campfires that are improperly tended or not completely extinguished*
- *Children playing with matches*
- *Refuse or camp fires that lift burning embers up into the air to fall on dry grass*
- *Glowing carbon particles blown out of automobile exhaust pipes*
- *Superheated catalytic converters beneath automobiles driven over dry grass*
- Hot boxes *that develop on railroad car wheels*
- *Dry grass blowing against electric fences*
- *Downed power lines*
- *Fireworks*
- *Uncleaned roof gutters harboring dry leaves into which sparks may fall*
- Self-combustion *caused by oily rags or by bacteria in damp hay stacks*
- Pyromaniacs—*those mentally deranged people who get their kicks out of setting and watching fires burn.*

Fighting a Grass Fire

Let's take a look at fighting a wildland grass fire on a windless day. Caught early, this might be considered one of the simplest fires to fight. If on level ground and discovered in its incipient stage, a simple attack may be to drive around the periphery of the fire with an engine, or perhaps with a pickup truck carrying a pump and fifty or more gallons of water. A 1- or 1½-inch hose played on the fire by one or two firefighters walking ahead of the engine or pickup should be able to effectively contain such a fire. A few people using backpack extinguishers, swinging beaters, or wet sacks might also be able to put out small grass fires.

If the fire develops into a large grass or brush fire, an updraft of heated gases produces an inward flow of air and burning embers toward the fire center, even on windless days, feeding it oxygen, accelerating its burning. Such an inward air flow can increase the speed at which the fire burns outward by its blowing outer unburned grasses down over inner, already burning grasses.

On windless days with high humidity (70 percent or more), the burning rate of grasses may be only a couple of feet a minute, with flame temperatures of perhaps 1,600° F. On dry days with low humidity (30 percent or less), the burning rate may be 100 feet per minute or more, with flame temperatures of perhaps 1,800° F.

A danger at any wildland fire is the developing of *spot fires*. These often result in an unexpected extension of a nearly extinguished fire. Spot fires are caused by burning embers, called *firebrands*, being lifted up into the air by a wind or by the fire updraft; they may be blown laterally to some nearby dry fuel area. The direction taken by the smoke of a fire tells firefighters where they should go with backpack extinguishers to look for spot fires. If a neighbor's home upwind of you is on fire, even in a city, take precautions that any firebrands dropping on your home or property are immediately extinguished. Here is where your emergency home fire hose can come into use.

A wildland fire may rekindle if the ground has a layer of dead litter through which grass had been growing. Such litter may have been soaked by water that put out the grass

fire, but a still-hot coal may develop flames again when the water dries out of the litter above it.

Fighting Mixed Brush and Grass Fires

What was true of grassland fire fighting is also true when fighting mixed brush and grass fires. There is the added factor of the longer time required to consume brush branches because of their smaller area-to-fuel-volume and greater water content. A dry grass stalk is a thin-walled, ignitable hollow tube with a large area-to-fuel-volume factor, allowing rapid oxidation or burning of the grass. In a few seconds, a stalk of dry grass may be chemically converted to hot gases and radiant heat energy. On the other hand, dry bush branches may take from several minutes to half an hour for most of them to be consumed by fire. The thicker and greener the branches are, the more heat and time required to dry the brush to the point where it supports combustion. The fire may smolder in heavier branches near the ground and in root systems for hours and even days.

One method used to confine a fire and prevent rekindling is to dig or scrape a 12- to-24-inch-wide path down to mineral earth (no growing material in it) all around the fire. This may be called a *fire line*, a *containment line*, or a *control line*. It should serve as a *firebreak*, or fire stopping area, when fire reaches it. Wider firebreaks may be cleared by a tractor with a scraping blade, or by bulldozers (*dozers*). Plows can also be used to develop fire lines if the area is not too rocky. A rule-of-thumb says that a control line on the flanks of a fire should be at least as wide as nearby fuel is high. When the firebreak is across the head of an approaching fire or above it on a hillside, it should be at least twice as wide.

Natural firebreaks, such as sandy areas, rock outcroppings, or areas of sparse fuels, make good *anchor points* or starting and ending points for fire lines. In remote or steep country, control lines are more often scratched out by hand, using McLeod, Pulaski tools, or shovels, Fig. 12-1.

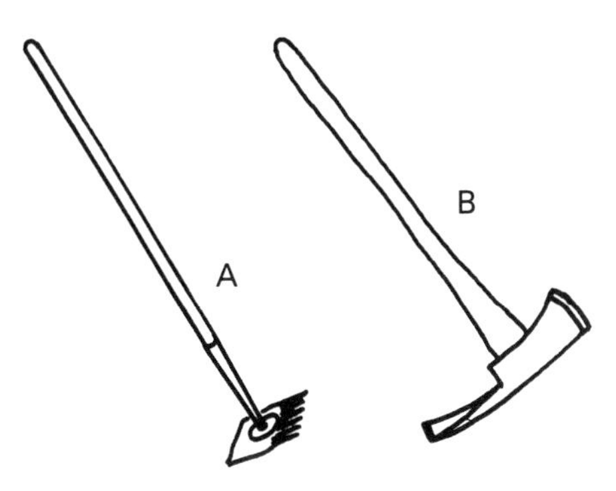

Figure 12-1. Wildland-fire hand tools. (a) A hoe-rake "McLeod." (b) An axe-adze "Pulaski."

Developing Fire Lines

Fire crews work on constructing fire lines in one of two ways, the passing or leapfrog method and the move-up method. In the leapfrog method, whenever a person completes the section being worked on, he or she moves ahead of all the others and starts working at the front of the line, about ten feet ahead of the previously leading crew member. In the move-up method, each member of the crew completes approximately a thirty-foot section of the line, and then all move forward together. Usually a crew consists of about five persons. Firefighters should work at least ten feet from anyone else to keep away from swinging, sharp-edged tools.

Two engines and crews may also operate in a leapfrog manner. The first engine makes a direct attack along one flank of perhaps a 100-yard fire line. The other engine starts 100 yards ahead of where the first started. When the first crew extinguishes its line, it will take up a position 100 yards ahead of the second engine's crew, and so on.

At least one piece of responding fire apparatus must remain at the scene of any small fire until mop-up is completed—until every wisp of smoke has been beaten out or watered down, and the fire is dead out! Brush branches

may appear to be out, but may smolder for a long time. Dry cattle droppings in a field require a great deal of mop-up work. They may have to be shoveled into a small pile and soaked with water, or even dunked into a bucket of water until dead out.

People living in rural areas have special precautions that they must take.

It should not be inferred that grassland fires on windless days may not be dangerous. If fires move into a narrow grassy valley bottom, the hot gases being developed produce an updraft. This can result in rapidly moving fires up one or both sides of the valley. Such fires, which usually involve some brush, can travel at high velocity and with a roaring sound that causes even the hardiest firefighter to take heed. The narrower the valley is, the greater the possible velocity of the updraft and speed of fire travel up the slopes. If the grass in front of a firebreak area can be wetted down, the fire will be slowed and is more likely to be stopped by a control line.

Once fire makes its way into a narrow valley bottom, it is probably best to try to establish a firebreak down and across the valley, ridge to ridge, ahead of it. Firebreaks cleared horizontally along a steep hillside of grass and brush must be quite wide if they are expected to stop rapidly ascending fires. If there is a structure on the hillside, a wide firebreak must be cleared at least thirty feet below the structure with somewhat narrower ones up and down both sides and above the structure. Fires burning over the top of a ridge and starting downward on the other side can often be stopped by a relatively narrow fire line dug down to mineral earth.

Anyone living in hilly or wildland areas should try to keep an adequate cleared area around home and buildings, perhaps a hundred feet for the areas below the structures. Roof coverings should be fireproof, not wooden. There are bushes, plants, and gardens that are considered to be fire retardant. Some of these might be ice plants, cactus, and rock gardens. There have been cases where such bushes planted around a home during Santana wind fire conditions made saving the home possible. The fires burned up to the fire retardant areas and were held there.

Whenever a grass fire starts to run up toward a timber line, it will be necessary to develop a firebreak as far below the timber as possible to prevent a forest fire. A two- to five-foot firebreak manned with adequate personnel may be able to stop such a fire on windless days. But many factors must be considered. For one thing, in North America, the water content of fuels on north slopes may be markedly greater than on south slopes. Fires will usually burn slower on north slopes—faster on south slopes.

Early morning hours usually find descending cool air masses on slopes, but after the sun has heated a valley and slope, a warm upward breeze develops which can drive a fire uphill rapidly.

Sometimes sections of a grass fire will extinguish themselves before they reach a fire line. This may be caused by the fire on both sides being extinguished, greatly reducing the radiant energy striking these grasses from the sides. Sometimes the water content in some areas of grass is higher, slowing the burning. It is considered good fire procedure to ignite with a *fusee* (flare), a torch, etc., all unburned grass and brush between a control line and an extinguished fire's perimeter. Such *burning out* or *clear burning* produces a completely burned area up to any control line.

Small Grass Fires

First response to an alarm for a small grass fire on windless days might be only half a dozen firefighters, a single brush truck with perhaps 100 to 200 gallons of water, shovels, Pulaskis, backpack extinguishers, McLeods, and beaters. At least one back-up pumper and water tender must also respond as soon as possible, in case the fire begins to run out of control. If the fire involves more than an acre or so, the first response might be augmented by other pumpers, probably vehicles capable of operating on rough terrain while in motion. How much manpower and equipment will be required can only be determined by an experienced officer or firefighter at the fire ground. Should a wind suddenly spring up, a simple fire may become a roaring inferno. Suppression plans may have to be changed, and more help might have to be summoned in a hurry.

The wearing apparel for wildland firefighters may be the standard turn-out gear, or light Nomex pants and jacket, goggles, and some form of headgear with face protector. However, during hot weather, lighter fireproof jackets and pants may be issued. Rural volunteer help may show up with only shirts, jeans, boots, perhaps gloves, and some form of face-guard helmet. Often only work hats and perhaps dark glasses will be the only specialized parts of their uniforms. Such non-department people, if accepted as *temporary firefighters*, should be deputized by the officer in charge as temporary firefighters to allow them to be included under worker's compensation. Many departments will not accept unknown help at wildland fires.

What about grassland fires on windy days? In general, winds tend to appear around 10 a.m. and continue until sundown. These hours are known as the *burning period* and are generally considered as the most dangerous fire times of the day. On large wildland fires, containment may be reasonably effective during night hours, only to be lost in the late morning hours when the winds start again, and the sun warms and dries any dew that may have settled on the fuels during the night.

Burn Patterns

The spread of a wildland grass fire under various wind conditions follows fairly simple logic. On a flat plain and on a windless day the *burn pattern* would have a circular perimeter with the original ignition point in the center. When extinguished, you would see a round, black burn pattern.

If a strong easterly wind is blowing, what might be the shape of the burn pattern? It would probably be pear or teardrop shaped, the narrow starting point near the east end, with the *heel* of the burn pattern a little east of the ignition point, and the wider *head* at the far west end. Burning ember spot fires blown ahead of the fire may be found out west of the head. If the wind becomes variable, shifting sometimes toward the north and sometimes to the south, it will produce a greatly widened head pattern.

There are various conditions in wildland areas that determine when the most danger exists for fire fighting.

When the starting point of a fire is on a slope, with no wind, can you see how the burn pattern might be an elongated oval with the wide end of the pear-shape up the slope? If a wind were coming from the left-hand side, the burn pattern would widen to the right as the fire ascended the slope, forming a rough triangle. What might the burn pattern look like on a windless day if the starting point is at the base below the head of a narrow ridge? How about a teardrop shape with the narrowest area at the starting point and spreading out as it ascends the ridge on both sides?

Why might understanding burn patterns be important? Well, if an officer can guess the burn pattern a fire will make, it tells him where to place equipment and deploy firefighters to develop effective fire lines to stop the spread of the fire.

Attacking Wildland Fires

A basic *direct* attack on a grass fire with some wind blowing may be to station the first-in and second-in pumpers at the heel of the fire. They could start a *progressive hose lay* of 1½-inch supply-line hoses, with water-thief fittings every 200 feet along both flanks. To each water-thief outlet is connected 100 feet of 1-inch, single-jacketed hose with gated nozzles, Fig. 12-2. The fire at the heel could probably be put out with the first 1-inch lines, then the fire would be progressively attacked along both flanks. While the 1½-inch hoses may be used to fight the fire as the lay progresses, the many smaller 1-inch hoses are quite effective at knocking down and mop-up, as well as protection of the 1½-inch hoses (should they be imperiled by slop-overs or spot fires).

When it is required to extend the 1½-inch line, a *hose clamp* is applied across the hose a few feet behind the nozzle. The nozzle is removed, and a water-thief fitting is substituted. A new 200-foot section of 1½-inch hose

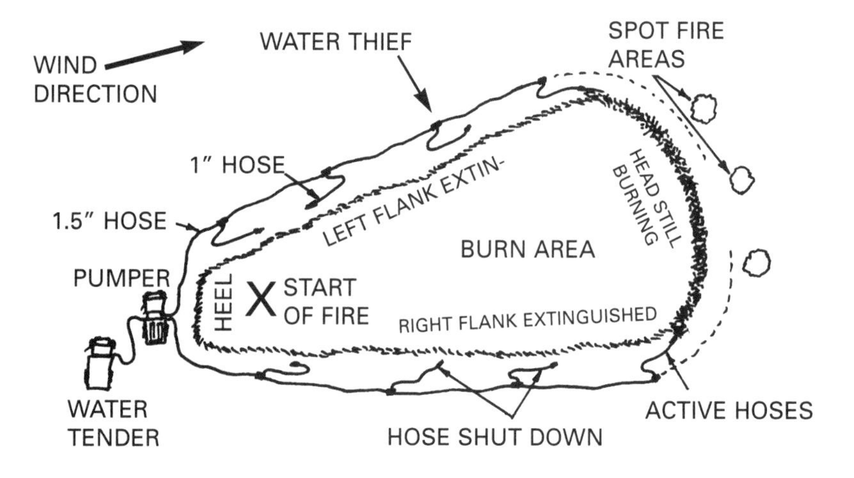

Figure 12-2. Burn pattern with wind blowing and possible progressive hose lay with 1″ and 1.5″ hoses from a pumper with water tender.

Wildland firefighters now carry portable fire shelters on their belts.

and another 1-inch nozzled hose are coupled to the water thief. The 1½-inch nozzle is then screwed onto the newly added 1½-inch hose before the hose clamp is removed. On a windless day, the two progressive hose lays and their firefighters should meet halfway around the perimeter of a now-extinguished fire.

If there is a significant amount of wind, a third pumper may start at the head of the right flank. The firefighters could start out across the head with a progressive hose lay and meet the left flank attackers on the left side of the head about when the right flank attackers reach the third pumper position. There are many other methods of attacking fires.

Local wind conditions may change any basic attack plan. For example, surface fires under dry fuel conditions may burn only about 100 feet per hour into a wind, but in a down-wind direction, driven by a strong wind, they may exceed 35 mph, faster than an engine can travel over rough ground, which can mean real trouble.

Sometimes it is possible to fight grass and perhaps sparse brush fires from inside the *black*, or burned area. Once burned, a black area should not burn again. In this area the firefighters will be safe in case of a sudden wind shift. While the black may be warm on the feet through boots, and will dirty the hoses and possibly even burn them, it may be a place of safety in an emergency. It is sometimes advantageous to detail a pumper into a cooled black area to allow fighting a flank or head from inside the fire perimeter as well as from outside it. This might be advisable if a fire is approaching a brushy area or timber where it could be dangerous to send a crew ahead of the fire. In heavy-brush fires this may not be possible because such fuels take a long time to burn out.

If trapped in a wildland fire, firefighters should look for rock outcroppings, cleared areas, or sandy spots. They may have to chop out brush to form a clearing and throw the cut brush downwind and try to dig a hole into which to crawl! This is where one of the *portable fire shelters* carried by wildland firefighters comes in handy. These consist of an aluminized piece of non-flammable mylar-coated woven fiber material of about 7- x 3-foot dimensions, with little handles on the inside surfaces. The firefighter finds or scratches out a patch of cleared ground, lies down, and pulls the shelter over himself or herself. While not too effective against direct flames, the shiny surface of such a shelter is quite effective against radiant heat. The shelters fold up into small packages that can be carried on the firefighter's belt. They offer protection for those they cover from temperatures up to nearly 2,000° F.

Backfires are fires purposely set to burn backward into the wind and toward an approaching wildland fire head to produce a burned-over area which should stop the advancing fire. Backfires are often used to protect structures in wildland areas, or to protect a forest or brushy lands. A road is often used as a fireline for backfires. There may be great danger in setting backfires. Before starting one, a person should be sure that he or she can stop it if necessary. If the wind should suddenly shift 90° to 180°, a backfire may only add to the size of the whole fire. Usually only trained officers should order backfires to be set. Backfires may burn into a light breeze, but any sudden increase in wind velocity may develop spot fires back over the point where the backfire started.

A brush fire is more concentrated and intense than a grass fire, particularly later in the summer. It can produce violent updrafts over the center of the burning area and a high probability of spot fires. Spot fires in bushes cannot be stamped or beaten out as can often be done with grass fires. They require an application of water, or the uprooting of the whole bush and throwing it into a burned-over area to let it burn itself out. If a burning brush patch is large, it may have to be surrounded with a firebreak.

A point of caution if fighting wildland or any other fires where poison oak or poison ivy may be burning: The smoke these plants give off will be carrying their poisonous oils. If such smoke is allowed to coat the skin or eyes, or worse yet, if it is inhaled, it may cause great irritation to the affected parts, if not death. Keep away from all smoke suspected to be coming from these burning plants. Goggles and a handkerchief over the nose and mouth may give some protection. It is far safer to wear self-contained breathing apparatus with a mask. Button shirt collars, wear long sleeves with buttoned cuffs, and gloves. Be careful when removing clothing later because any residual oils will be toxic to your skin. Shower with soap as soon as possible.

Again, all structures in wildland areas should have all dry grass and brush cleared away from them for a minimum of 30 feet (100 feet is much better) in all directions. Structures nestled under beautiful forest trees will probably be lost. Fireproof shingles or tile roofs may prevent the roofs from burning, but radiation alone from a close-by forest fire, or even from a nearby building on fire, may ignite wooden exterior walls of a home or other structure. Even if the walls are fireproof, the interior areas may ignite due to heat radiating through the glass of the windows. If the radiant energy of the fire doesn't get the structure, then falling burning branches or trees may. It's not a happy thought.

The smoke from burning poison oak or ivy can cause great irritation to skin, eyes, and lungs. Keep away from it!

Structures in brushy areas can sometimes be saved if an engine with a couple of 1½-inch hoses can direct an adequate amount of water toward the center of an approaching wall of fire. The water should be in the form of spray or fog. The approaching fire should be pushed around both sides of such a spray bank and around the structure to be protected. The walls and maybe the roof may have to be doused from time to time to prevent them from bursting into flames due to radiation from the nearby fire. If radiation overheats the firefighters, they may be able to shift to a fog attack for a time to absorb much of the radiant energy coming toward them from the approaching fire.

Because of the intensity of the heat in brush fires and the height of some bushes, firebreaks along the flanks and heads of such fires must be several times the width of those for grass fires. This is where the bulldozer comes into use. It can cut a containment line eight to fifteen feet wide in a very short time compared to what a fire crew can do. It is important that the dozer operator throw all of the uprooted dirt and bushes to the side of the firebreak away from the fire so that the approaching fire does not find a high stack of flammable material in its path to the firebreak.

When control lines are cut laterally along slopes below a line of fire, they are called *undercut lines*. They are subject to slop-over from rolling matter, such as round embers, burning logs, or burning pine cones. The lower edge of an undercut line can be built up with a *berm*, a little ridge made up of stones and dirt, to make the undercut lines more effective at stopping rolling firebrands.

A brush fire that has been driven southward by a north wind for several hours develops a well-defined head and flanks. If the wind

suddenly shifts and comes from the east, the original head becomes the left flank, and the original right flank becomes the head. Crews working on the original right flank now find themselves fighting a rolling head fire instead of a less dangerous, slower burning flank fire. If the fire line that the crew constructed is not wide enough to stop the fire head, they are in an untenable position and must retreat to safer areas, perhaps to previously cleared *safety islands* or to cooled black areas. Here is where hand-held radio transceivers and helicopters can provide life-saving rescue missions, if necessary. Meteorological information available at the incident command (IC) post or base in large fires should be able to give forewarning of such a shift in wind direction, allowing the crews to be told to evacuate their positions, or to make some other arrangements.

As soon as all fire lines are holding and no spot fires are occurring, mop-up can begin.

Mop-up

When the whole fire perimeter is finally contained, with all fire lines holding and no spot fires being developed, mop-up starts. This consists of working in from the perimeter, making sure that all fuels still smoldering, which might re-ignite later, are extinguished. Old, dead trees, or partially burned trees near the perimeter of the burn, if still smoking, must be felled to prevent burning branches and bark from being wind-driven across any nearby control line. Smoldering logs should be turned to point uphill, if possible, or be locked into position with rocks to prevent their rolling. Islands of brush that have been isolated by the fire, but which might ignite later may be torn out, piled up, and burned far inside the perimeter. Mop-up may take a very long time.

Forest Fires

Wildland burning that occurs between the ground and the lower branches of forest trees, involving grass, brush, or surface litter (called *duff*), is known as a *surface fire.*

Roots and subsurface rotting organic matter, called *peat* or *humus*, may smolder for long periods of time, showing little or no smoke. Such burning of subsurface peat and root smoldering is known as a *ground fire.* Ground fires have been responsible for the rekindling of fires days, weeks, and even months after an area has burned over. Ground fires have been known to burn in peat or roots under roads that had acted as firebreaks in a previous fire. They suddenly rekindle into a fire months later on the other side of the road. Aspen trees, for example, send out their roots for a hundred feet or more just under the surface of the ground.

Ground fires in forests have been made more dangerous by past ideas that all forest fires should be extinguished. It is now known that *prescribed* forest fires should be used. These fires may be deliberately set to burn the underbrush and smaller trees as well as lower branches of large trees in order that wildland fires will burn only the grass and brush and not envelop the larger trees. In this way a more natural forested area will be produced. Because of the denseness and the undergrowth in many of our present younger forests, it may be better to just let them burn and start over with a new, better managed, more ecologically natural forest.

Forest fires that move up from surface fuels and finally ignite the upper branches of trees are known as *crown fires.* When forest fires begin crowning, it indicates intense surface heat and trouble for the firefighters. Crown fires can be driven at high speeds by winds and are out of reach of all fire-fighting equipment except aircraft. With wind-driven fires, an attack can be planned, but when winds drop off, a *plume-dominated* fire may develop. Its rising heat develops extremely active fire in tall trees, and such fires may take off in any direction, presenting extreme danger to firefighters and difficulties for fire-fighting planners.

Aerial Fire Fighting

Aircraft have been used successfully in wildland fire fighting since the mid-fifties.

Small and medium ex-bomber planes, known as *air tankers*, drop water, or water mixed with chemicals, either on slow burning fires, or just in front of the heads of more intense fires. One of the first fire retardant drops was a sodium-calcium-borate solution, which is the reason why modern fire retardant planes may still be called *borate bombers*. The original solutions were only really effective as long as the retardant drop remained wet. Unfortunately, the borates sterilized the soil for a year or more and were abrasive to pumps. The modern fire retardants are chemicals that retain their retardant capabilities after they have dried, then act like a dry powder extinguisher agent. They are *MAP (*from *monoammonium phosphate), DAP (diammonium phosphate)*, or may be *ammonium sulfate*, with thickening and sticking agents added. These are actually commercial fertilizers that happen to have good fire retarding properties. Those areas where retardant drops were made during a summer or fall come up greener faster than surrounding areas the next spring.

Air tankers are operated by the U.S. Forest Service, by state forest services, and by private companies under contracts. The airplanes they use range from small bombers capable of handling a few hundred gallons of fire retardant, to medium and larger bombers capable of carrying 1,000 to 3,000 gallons. Many military bombers, for which there is little use with the ending of the cold war, are being converted to air tankers and are able to deliver considerably more than 3,000 gallons of retardant at a fire scene.

When attacking a fire, a bomber, unless it is being directed by an air attack coordinator or drop commander in a scout plane, usually makes a first pass over the fire ground it is going to hit to determine the conditions that exist. It then banks around and comes in again, flying low over the terrain and usually from the same direction as the first pass. Just before it reaches the target area, at an altitude of about 125 feet above the terrain, it opens its tanks, and the whole load may be dropped in a second or two. With multiple tanks, one after the other can be opened to produce a longer line of retardant on, or ahead of, the fire. As the liquid falls, it breaks up into droplets that land on the grass, brush, or trees as a thick red-dyed slurry. While wet, it cools the fuels it hits, and when dried, it interrupts the chemical ionization that occurs during oxidation, greatly retarding the chemical reaction of burning.

Airplanes used to drop borate solutions on wildfires. Although better retardants have been developed, the planes may still be known as borate bombers.

An aerial drop, sometimes called a *pre-treatment*, may result in the immediate extinction of low-intensity fires. Since an aerial spray does not cover all fuel in an area, it cannot be counted on to extinguish most fires. Ground crews usually have to follow up with a finishing attack.

When firefighters see a plane making its first pass over them, they had better move out of the way, or they will be hit with the red, sticky, stinging solution when the retardant is dropped by the plane on its next pass. The solutions are dyed red to allow already sprayed areas to be identified from the air. Retardants are non-toxic and should cause no lasting harm to skin, but they do sting if they hit the skin.

An 800-gallon drop can lay a line of retardant about 400-feet long (about 2 gallons per foot) and up to 20-feet wide. A 1,500-gallon drop can cover an area almost the size of a football field.

The pattern of an aerial drop is determined by how fast the tanks are opened, the altitude of the drop, and the wind conditions. A slow opening and low altitude produces a long thin line of retardant. The higher the altitude, the wider the line will be. An aerial attack on a wildland fire may consist of from one to half a dozen planes. They follow one another, laying down lines of retardants on or in front of the head of the fire. The planes must fly back to the airport from which they received their retardant supply, which may be many miles.

Fire retardants from the air can be effective on grass and many brush fires, but they have little effect on the rolling flames at the

head of a large timber fire. They may have a desirable slowing effect along burning flanks and may be used effectively in widening firebreaks ahead of large wildfires.

A prime requirement in fighting forest fires is the rapid determination of smoke plumes while they are still small.

Airplanes and helicopters are also used for reconnaissance of large fires. Infrared (IR) frequency cameras in planes can take pictures that will pinpoint the area on the fire ground that is producing the greatest radiation. This is probably the head of the fire. Large fires may branch out up different valleys, producing multiple heads. IR cameras also show where spot fires are, even if they are covered by heavy smoke. In the mid-1980s, amateur IR TV pictures were taken from reconnaissance planes and transmitted by radio to the IC base that gave the personnel below a real-time view of what the fire looked like from above. This was a great help in planning what should be done next. Commercial IR TV equipment is now used on many big wildland fires.

Helicopters

Helicopters have a variety of uses at wildland fires. Besides reconnaissance, they can be used to lay hose in remote areas. They can discharge water into open-topped portable tanks or sumps that have been airlifted to open areas near the fire. They can fly crews and supplies into and out of remote bases. They can make rescues or fly out injured personnel. They can start backfires by dropping chemical or jellied napalm *golf balls*, also called *helitorches*, which break into flames thirty seconds after hitting the ground. In large forest fires, hand-held, or even aircraft-deployed, flame throwers may be used to start backfires. Helicopters with 100- to 1,000-gallon metal buckets suspended below them can hover over a lake, fill their buckets, then fly to the fire where the water is dropped. Fire engines operating from remotely placed fire hydrants or from other sources of water can often be used to pump water into water-carrying helicopters. Dozens of large helicopters may be used around the clock during daylight hours in large wildland fires.

Spotting Fires

The danger of tinder-dry fuels in forests is often due to dead trees, standing or downed. Insects and blights kill millions of trees yearly, furnishing great quantities of dry, highly flammable fuel. When loggers fell trees and trim the branches from the logs, their slash is often left to rot on the ground, forming a widespread source of dangerous fuel for fires.

Lookout towers on high peaks are scattered around our forests. They are manned by spotters who are in communication with each other and the forest service headquarters by radio or telephone, or both. When a suspicious plume of smoke is spotted, its location can be accurately determined by visual bearings taken on it by two lookout stations. Suppose one lookout sees a plume of smoke in a direction 45° east of north, and another lookout sees it at an angle 270° from north. If a line is drawn through the first lookout's position on a map at a 45° angle from north, and a second line is drawn through the second lookout's position at an angle of 270° from north, where the two bearing lines cross on the map is the location of the fire. A fairly accurate position can be obtained by a single lookout's bearing if nearby ridges are identifiable.

If an observed smoke plume appears to have a small point of origin, a ranger in a four-wheel drive vehicle may be sent to investigate, provided a road runs near the location. Aircraft may be sent if there are no nearby roads. If it is determined to be an unfriendly fire (not a well-tended camp fire) that cannot be knocked down easily, this information is passed to ranger headquarters, and preparations for a battle begins with the mobilization of officers, firefighter crews, and equipment.

Fighting Forest Fires

It is in the incipient or surface fire phase that an effective water attack can be mounted against a forest fire. There is hope that it can be contained by a direct attack as long as it holds as a ground or surface fire. Crews can be brought in with shovels, backpack pumps, various tools, and mechanized equipment to attack it. If the terrain permits, pumpers and brush trucks can be used. Aerial tankers can be flown in to slow the fire before it becomes a runaway conflagration. In steep canyons, or with winds over 30 mph, or under heavy smoke conditions, air tankers may not be operational. Once the fire starts to move up into the branches and crowns, it may be termed a *canopy fire* and is out of reach of the ground crews.

In big forest fires, crews of up to twenty members will be detailed to a direct attack on the flanks. Natural flank edges are often formed by roads and fire trails. Strenuous attempts are made to hold such fire lines. When the head cannot be held by direct attack, a decision must be made where to try to stop the progress of the head, usually along some designated natural barrier, such as a rocky ridge. Foam may be spread ahead of advancing fire heads if fire engines can be brought in close enough. All operations are parts of a master *Incident Command System (ICS)* action strategy, even as the prongs of the fire fronts spread out along valleys, up canyons, and over ridges.

Fire jumpers, specially trained firefighters, may be air-dropped by helicopters or bombers into forest fire areas. There may be two to twenty firefighters making these jumps. They wear padded clothing for protection when they land. They may carry up to eighty pounds of equipment on their backs. They will have chain saws, hand tools, and possibly dynamite. With their handy-talkies, they will report fire conditions or request that additional equipment be dropped in to them. They may fell trees away from an approaching fire. Firebreaks may be developed and backfires set. Much of their fire suppression is by shoveling dirt onto burning downed trees or brush. Usually the jumpers must walk out to some pick-up point after the fire is controlled—with all of their equipment on their backs.

Incident Command Systems

When it looks like a fire may be, or is getting, out of control, or is starting to crown and to be more than a surface fire, plans are developed to set up an Incident Command System. Such a base or post camp will be set up somewhere near the heel of the fire. As the fire grows in area, the IC base will also grow until it may appear much like a small town.

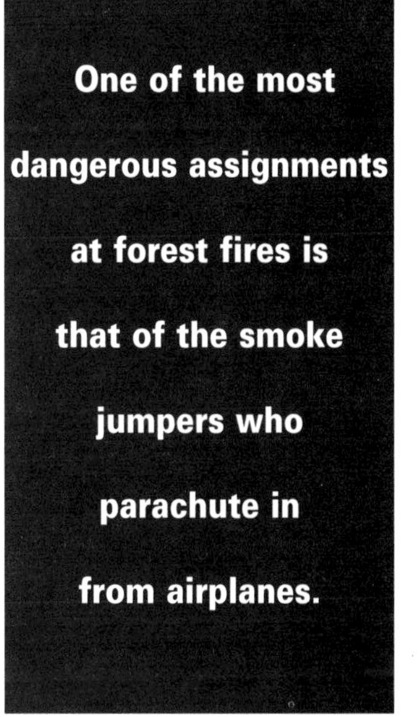

When you see a huge forest fire on a TV news program, what you do not usually see is the Incident Command System that is at work somewhere nearby. To give you an idea of the complexity of such a fire camp, the following are abbreviated outlines of what an ICS post might eventually consist.

When the fire is large, the ICS camp officers will be: *incident commander*, or *IC*, in charge of all fire activities; *information officer*, who checks and reports any safety violations to the IC; *safety officer*, who oversees safe working conditions for crews; *liaison* officer, who acts as liaison person with all of the mutual aid fire agencies; *agency representatives* from each fire agency assigned to the incident.

Under these ICS command officers are four separate sections:

1. ***Operations Section.***
Under its *Section Chief* are: a *Branch Director*, a *Staging Area Manager*, *Strike Teams*, *Task Forces*, an *Air Operations Branch*, and *Director*. Also there are: a Fixed-wing *Air Attack Supervisor*, a *Helicopter Coordinator*, an *Air Tanker Coordinator*, an *Air Support Supervisor*, a *Helibase Manager*, a *Mixmaster* who oversees mixing of fire retardants, a *Helispot Manager*, a *Deck Coordinator*, a *Loadmaster* for personnel and cargo, a *Parking Tender*, a *Takeoff and Landing Controller*, *Helibase Radio Operators*, and a *Helicopter Timekeeper*. These are the

people who will be actively working at putting out the fire.

2. *Planning Section.*

Equally important at a large fire, it has: a *Section Chief*, a *Resources Unit*, a *Leader*, a *Check-in Recorder*, a *Situation Unit Leader*, a *Fire Display Processor*, a *Field Observer*, a *Weather Observer*, a *Documentation Unit Leader*, a *Demobilization Unit Leader*, *Technical Specialists*, a *Fire Behavior Specialist*, a *Water Resources Specialist*, an *Environmental Specialist*, a *Resource Use Specialist*, and a *Training Specialist*.

3. *Logistics Section.*

It has: a *Section Chief*, a *Service Branch Director*, a *Communications Unit Leader*, an *Incident Dispatcher*, a *Medical Unit Leader*, a *Food Unit Leader*, a *Support Branch Director*, a *Supply Unit Leader*, an *Ordering Manager*, a *Receiving and Distribution Manager*, a *Tool and Equipment Specialist*, a *Facilities Unit Leader*, a *Facilities Maintenance Specialist*, a *Security Manager*, a *Base Manager*, a *Camp Manager*, a *Ground Support Unit Leader*, and an *Equipment Manager*. This section works closely with the Operations Section.

4. *Finance Section.*

It consists of: a *Section Chief*, a *Time Unit Leader*, an *Equipment Time Recorder*, a *Personnel Time Recorder*, a *Commissary Manager*, a *Procurement Unit Leader*, a *Compensation for Injury Specialist*, a *Compensation/Claims Unit Leader*, a *Claims Specialist*, and a *Cost Unit Leader*.

Descriptions of all of these jobs would be voluminous!

Not at the scene of the fire are many other participants, such as deputy sheriffs and highway patrol officers to control traffic and provide needed protection, headquarters administrators to make decisions concerning commitment of personnel and equipment in the event of multiple fires, emergency service communication personnel, relief agencies, plus other volunteers, such as trained Red Cross workers and amateur radio communicators.

Obviously all of these people would not be required at a ten-acre fire. But as the fire grows into the hundreds to tens of thousands of acres, and the time required to extinguish it lengthens, more and more of these activities and people will have to be mobilized.

When there is a large fire involving several buildings or blocks in a city requiring fire units from many fire stations, an Incident Command System operation may also be set up. Many departments today use an ICS format even for smaller fires. So may a large man-hunt involving many city police, deputy sheriffs, state police, and highway patrol units. Flood, hurricane, tornado, earthquake, or other disaster situations may develop an ICS. Of course, they will use only those ICS personnel necessary to do the job at hand. In all cases the officer in charge on any shift is the incident commander. The IC may be a chief, a lieutenant, a captain, depending on the size of the incident. In some areas, different nomenclature may be used, but in general the operations will be quite similar to those outlined here.

Radio communications by many involved agencies continually feed the latest information into the IC base on fire progress, placement of personnel, weather, position of mechanized equipment, needs for food and shelter, effectiveness of air drops, of jumper activities, reports from lookout towers, requests for first aid and ambulances, how many acres or square miles have been burned, where spot fires are occurring, and so on and on. Fire equipment at large forest fires may include U.S. Forestry units, Bureau of Land Management (BLM) units, state forestry units, county units, and often nearby city units.

Effects of Forest Fires

Wherever fire involves conifers and other pitch-bearing trees, the heat released may double. Gigantic clouds of smoke billow up from the fire storm. Fire whirlwinds develop small tornadoes, lifting large chunks of burning debris thousands of feet into the air. Where firebrands land, new fires start. In this way, fires jump wide firebreaks, such as highways and roads. The extreme heat of such a fire storm may drive fire down a hill almost as fast as it would burn uphill. Some pine cones open and burn easily. Some *serotinous* pine cones remain tight units in a tree but will drop down during a fire. The heat opens them to mature their seeds so that the next year they sprout to re-seed the forest.

Too often, wildland fires will be blown into towns or cities, enveloping them disastrously. At least in such areas, there is usually a good water supply with which to try to protect local buildings by keeping their roofs and sides wet.

As a raging wildfire approaches one of the natural barriers selected by the IC officers, it probably is burning up a sparsely wooded slope, with a natural or man-made firebreak at the top. Liberal dumping of fire retardants slows the burning as the fire rises out of the canyon. Crews *hot-spot* by clearing or pre-burning areas that might develop excessively hot fires when the head hits them. As the cool of evening approaches, the winds decrease, and the fire tends to settle down. If the plan has been successful, the line of fire is stopped, and mop-up operations can start. Some of the hot spots or prongs in other valleys may spring to life again and may develop into another conflagration the next day when the winds reappear. Eventually, the fires run into natural or man-made firebreaks and are contained. Rain often moves into a forest fire area and helps to slow or sometimes to put out

Mopping up after a forest fire may take almost ten times as long as was spent on extinguishing the fire.

In 1996, army tanks that had been declared "excess" by the federal government were first put into service for wildland fire fighting and rescue. Tanks can travel over difficult terrain carrying thousands of pounds of water, people, tools and equipment. Their wide, steel caterpillar tracks may alone put down, or at least slow, grass and small brush fires. They can travel through blazing areas and knock down reasonably large burning trees. Their slip-in tanks may carry 600+ gallons of water. With the water tank removed, they can act as ambulances since they then have room for several litters. Many personnel can be carried on the top surface in emergencies. They will go up and down 60 percent grades, cross three-foot ditches, and travel through forty inches of standing water. Tanks may be used to fight fire with hoses handled by firefighters outside, or if the radiant energy is too high, firefighters can fight the fire from inside a tank using a fire nozzle mounted in a rotary turret on its top. They can be carried to a fire scene on heavy-duty trailers. Once there, they have been found to be exceptionally valuable. They should become one of the most useful of all wildland fire fighting equipment in the future. Paid rides in them at firemen's musters should help amass considerable funds for rural departments. After all, who wouldn't like to ride in a real honest to goodness army tank?

the fire. In many cases, fires burning up a mountainside will suddenly be met with a change of wind direction—the downhill wind pushes the fire back on itself, and a natural backfire is developed, putting out the fire.

Even when the battle is won, mop-up and patrol must continue for weeks. Crews may come in from hundreds or even thousands of miles away. They will consist of federal, state, and local fire groups. Prisoners may be sent in to help fight the fires. Special fire-fighting groups may show up. All must be fed, paid, housed, and transported into the camp, around the fire, and home again. Control lines must be completed and burned clear. In some areas where fire lines existed, trained foresters *cold trail* the edge of the burns, feeling by hand whether all fire is out. Burning logs and stumps are extinguished, fire lines are clean-burned, and roads are cleared. It has been estimated that in many forest fires, as much as 90 percent of the time may be spent in mop-up and only about 10 percent in actual fire extinguishment. For grass fires these percentages might be more like 50/50. With brush fires, the mop-up depends on the size and quantity of the brush, number of trees, if any are involved, and the weather.

In all major fire disasters, the Red Cross will be prominent in the rear, the front, as well as in the background. They provide much needed food, drinking water, housing for those burned out of their own homes, and even mental health aid. Those who have gone through a catastrophic fire know just how important the Red Cross and their volunteer workers are to everyone.

We have taken a look at the basics of wildland fire fighting. But how about structure fires? Let's see first what the basics are of fighting fires in smaller homes and other single-story buildings and what we might do about them in our next chapter.

13 Single-Story Structure Fires

When the Fire Alarm Sounds

It is only natural when we think about extinguishing a fire that our concern is, "How would I put out a fire in my house." Let's assume at first that the involved house has only a single level. (The ramifications of a multistory building will be discussed in the next chapter.) Your fire might be only a smoldering couch, or a small kitchen fire, but it might be the total involvement of one or more rooms, or it might be an outbuilding. Your first action should be to alert the nearest fire department—dial 911—or whatever your local emergency number is.

The scenario for the fire department, after it is alerted by your 911 or other call will be something like this:

1. *The first alarm is sounded at the station. The fire fighters suit up, climb into a pumper or other apparatus, and start driving toward the given address with siren, yelper, horn, or bell sounding.*

2. *Size-up is made on the way: What is the weather? What is the smoke color and volume? Must adjacent buildings (exposures) be protected? Does it appear that additional help should be requested?*

3. *At the fire ground, is any rescue required? Should a second alarm be called in? If these added fire fighters and engines are not enough, will a third alarm have to be sounded? A fourth alarm?*

4. *What kind of attack is required to contain or prevent spread of the fire? Where should hoses be laid?*

5. *Extinguish the flames.*

6. *Overhaul or mop-up the fire.*

7. *Replace equipment on the mobile apparatus, return to the station, secure the vehicles, clean hoses if necessary.*

Everyone should plan how to extinguish a fire in his or her home

Let's take a closer look at these basic operations. When a fire station's alarm sounds, it may be by bell, siren, or horn at the fire house, or by radio. The firefighters put on their protective bunker or turnout clothing. The destination, kind of fire, and units to respond may be announced in several ways: By voice announcements over a public-address circuit in a manned firehouse; by voice announcement via radio after a transmission of alerting tones; by a tapped-out bell code; by voice announcement via radio to vehicles as soon as firefighters check into service; by information printed on a blackboard at the station. Whether the department is rural or urban, small or large, determines what system is used to advise firefighters where to go and what vehicles should respond.

The doors of the fire station are raised or swung open as the apparatus drivers start their motors. Two or more firefighters will join the driver on each engine before it rolls out. The address of the incident usually gives a clue as to the fire problem. Is it in a rural, residential, downtown, or industrial area? The first response, if in a city, might be either one or two engine companies (pumpers and crews) and perhaps a ladder-carrying apparatus (truck) company. In rural areas, one or more water tenders would normally follow the first-in engine or engines.

As soon as each vehicle is underway, it usually announces by radio its unit number and its service status.

Preplanning by officers and firefighters has previously determined the best probable

route to reach the general area of the announced destination. This planned route is followed by the drivers. If any extensive street work is being done, or if the fire occurs during heavy traffic periods, it may be necessary to modify the preplanned route. The first vehicle to spot smoke usually advises the others by radio what color the smoke is, its volume, its direction, and what involvement is seen.

Size-up of the situation is a mental evaluation made by the first-in officer, who considers the weather, time of day, quantity of immediate water needed, number of men responding, whether smoke is seen or not and its color, how straight and fast it rises (for wind conditions), and if flames are visible. If flames are seen, the officer reports back to HQ that it is a *working fire*. If flames are not visible, the officer begins to plan one way. If flames are seen, and if they are through the roof, the officer may change his or her attack plan.

At a Fire Scene

On arrival at the fire scene, any possible rescue of people must be the paramount consideration. If all of the occupants are outside, and the citizens on the scene report everyone has vacated the building, a search may be unnecessary. On the other hand, people don't always know for sure, and a search may have to be made. If calling elicits no internal response, and if no occupants or neighbors are outside a burning building, a search is normally indicated.

The most important first action for firefighters to take when arriving on the scene is to save lives.

Fires are most often caused by people, so it is quite possible that someone was, or is, in any building that is on fire. No less than two firefighters should put on breathing apparatus and together scout any building enveloped in smoke. In any totally involved room with smoke and flames down to the floor, there may be little hope of anyone surviving. All other rooms must be searched quickly for sleeping persons, invalids, or children. This includes under the beds or in closets where children often hide, as well as in the basement. The involved rooms may be attacked by firefighters not part of the searching crew. It is often necessary to detail a separate crew to knock down small areas of a fire to permit proper searching.

Any people and animals found in the house who are unconscious must be carried or dragged out of doors or windows. Once outside, if a person's breathing is labored, pure oxygen may be administered. If not breathing, mouth-to-mouth resuscitation should be given. (All citizens should take a CPR course from the Red Cross or other organization before emergencies occur.) An ambulance should be called. If a body is burned past all hope, it should be covered with a blanket, and the local police and coroner should be notified.

A check must be made to determine if the gas line is shut off and if the electrical main switch should be pulled. It may be possible that an electric current might flow down a straight stream of water to a nozzle and to a firefighter on a wet floor. At night, however, it might possibly help to leave electric lights on if it appears that straight streams will not be applied to electrical equipment. Decisions, decisions!

When arriving at a fire ground, the driver of an engine will usually become the engineer—the person to operate the pumping controls. The other firefighters begin to lay out hoses if there is smoke showing. If no smoke shows, two firefighters would probably grab fire extinguishers of different types and enter the house to determine what action should be taken. With a little smoke showing at a rural fire, the engine and water tender will usually go directly to the fire. The first-in firefighters should enter the building with portable fire extinguishers to check for any necessary rescues. If none, but the fire appears to be more than the extinguishers can handle, perhaps a 1-inch hard-line carrying water under high pressure may be pulled in first to see if the fire can be stopped with a small-volume, high-pressure fog attack. If the fire is too involved, at least two 1½-inch or larger lines will be deployed, preferably from two different directions, perhaps 45° apart, both toward the fire source.

It is important that a structure fire not be attacked by two crews moving 180° from each other. This might drive the fire into one or both of the crews.

At a rural fire, a 2½-inch or larger hose can be coupled from a water tender to the pumper to provide more water for the fire than is available in the pumper's tank. If the fire shows no sign of knocking down within one or two minutes, it would be prudent to call for back-up and more water tenders from mutual aid sources.

Hose Lays

If smoke is showing in urban fires, a *forward hose lay* may be executed (Fig. 13-1). As the engine approaches a fire scene, one possibility is to have it stop at the hydrant closest to the fire. A *hydrant operator* and one end of a 2½-5-inch supply line hose, with perhaps a four-way valve attached to it, may be dropped off. The hydrant operator anchors the hose to the hydrant and signals the engine to proceed to the fire scene. The hose is always carried prepacked with its 50- or 100-foot lengths coupled together in the hose-bed in the back of the pumper. The hose pulls out onto the roadway as the engine moves toward the fire. As soon as the engine reaches its destination, the hydrant operator couples the hose to the hydrant outlet (a four-way valve added between hydrant and hose would allow other engines to couple their hoses to the same hydrant later).

When the engine arrives at the fire scene, firefighters will couple the large supply hose from the hydrant to the pumper's water inlet. At the same time, other members of the crew lay a 2½-inch (or larger) hose line to the fire ground, terminated with a double-gated wye to feed two 1½-inch hoses. (Larger hoses are used at large fires.)

The engineer couples the water pumps to the pumper's engine to first supply the tank water to the hose lines, and later to relay city hydrant water to the fire. The pumps increase the relatively low-pressure/high-volume hydrant water to the fire lines.

As soon as the hydrant line is attached to the pumper inlet, the hydrant operator is signaled to turn on the water. A small drain valve on the side of the pumper's suction inlet may be opened to *bleed* the air that was in the empty hose as hydrant water begins to flow toward

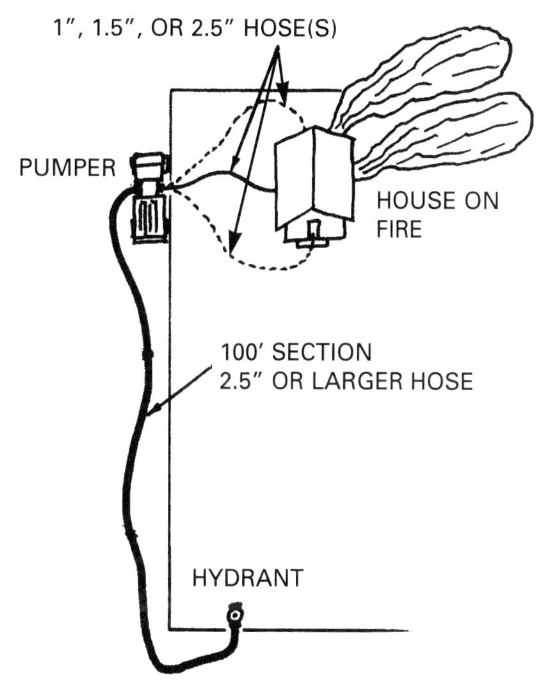

Figure 13-1. Possible hose lay from a hydrant to a fire scene.

the pumper through the hose. When hydrant water fills the supply hose, the engineer closes the bleeder valve and opens the suction inlet valve. This substitutes hydrant water for the engine's own tank water, which has been feeding the pump and the hoses up to this time.

While water is being pumped to the fire scene, the tank filling valve may be opened slightly to allow the pumper's own tank to be refilling. This will assure water availability in the event of some kind of a failure of the hydrant water supply.

Another possible method of delivering hydrant water to the fire scene is to make a *reverse hose lay* (Fig. 13-2). In such an operation, the engine proceeds to the fire scene first, dropping off an officer and one or more firefighters. They remove from the engine's hose bed enough 2½-inch and 1½-half-inch hose to

reach into and around the building on fire, a manifold or gated wyes, plus ladders to reach the roof. The hose is anchored at the fire scene and as the engine drives to the closest hydrant,

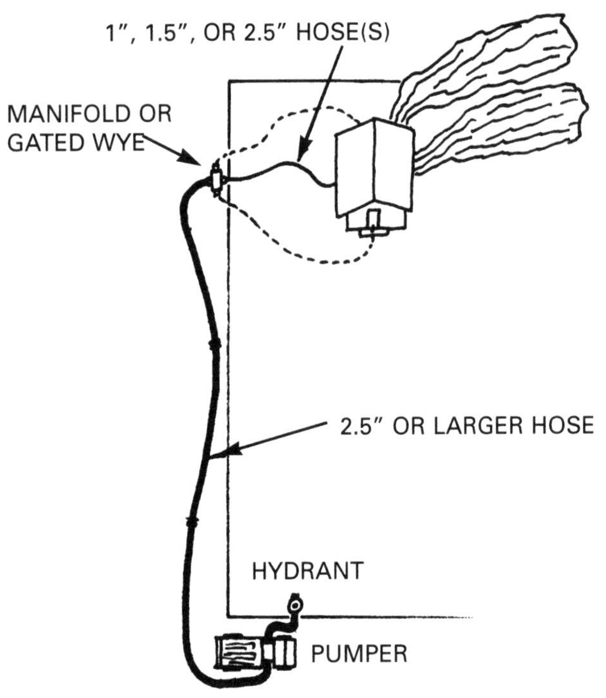

Figure 13-2. Possible reverse hose lay from scene to hydrant.

the hose pays out (is pulled off of the engine's hose bed) onto the roadway. When the pumper reaches the hydrant, it parks close to it. A large diameter soft-suction hose is coupled to the hydrant outlet and to the pumpers's water inlet pipe. The hydrant valve is turned on, and the engine pump is put in gear, pumping water to its gated outlets, which are still closed off.

The hose that was pulled out of the engine's bed while moving from the fire scene is disconnected at one of its screwed-together connections. It is coupled to one of the water discharge outlets on the sides of the pumper. The outlet valve is opened, and the engine's throttle is advanced as necessary to raise pump pressure to force water to the fire scene at the desired pressure.

At a rural fire scene, if taking water from a pool, a 2½-inch or larger hard suction hose, with a strainer attached to one end, is substituted for the soft-suction hose that is used from a hydrant to the pumper. (Soft-suction hose would be collapsed by the suction placed on it by the pump.) The strainer is placed at least two feet below the water surface. The water-tank outlet valve is opened to furnish priming water for the centrifugal pump. When the suction valve is opened, water is drawn up into the pump and to the pumper's outlet pipes.

If two engines respond to an alarm, the first engine may proceed directly to the fire and begin fire fighting with water from its tank. The second engine may make a forward or reverse hose lay between the first engine and the hydrant.

If a fire appears to be under control when a second engine arrives, it may only aid the first with hoses, ladders, venting, and other necessary activities. If the fire continues out of control, the second engine may make a second hose lay for itself from the hydrant and add its water and crew to the attack. If the fire warrants it, the IC commander may call for a second alarm which normally elicits the response of a second pair of engines.

If the fire involves only a small area in a kitchen, for example, an application from a portable CO_2, or a dry powder, or other fire extinguisher may be all that is required for containment and extinguishment. While CO_2 extinguishers may be the least messy, a shot or two of high-pressure fog from a ¾- or 1-inch high-pressure hose line may blow out and cool a small fire in a kitchen and produce very little water damage. This would be followed by the opening of all upwind and downwind windows and doors to vent the smoke to the outside.

Chimneys may gather a coating of grease, soot, and creosote on the inner surface, which can catch fire. If it is a modern ceramic flue chimney, in some cases it may be better to let the coating burn itself out. If the chimney is old or has a metal liner, the fire should be extinguished, and then all areas around the chimney should be checked thoroughly for heat build-up due to loose or missing mortar or bricks.

If it is decided to use water to extinguish a chimney fire, it is usually best to direct fog or CO_2 up the chimney from below. If any kind of stream is directed down a chimney, it blows

soot and perhaps firebrands out into the room below. If such an attack must be used, the front of the fireplace or stove should be covered with a tarpaulin to protect furnishings in the room before directing any water down the chimney.

If a room is only partially involved, the flames can probably be knocked down within a few seconds by a high-pressure, 30° fog attack. Fire containment and extinguishment may occur almost simultaneously.

Often, with small fires in dwellings, all that may be needed is a fire extinguisher, or perhaps a 1-inch, high-pressure booster line fog attack. A back-up for any high-pressure, 1-inch line application might be a 1½-inch hose with a combination solid-stream/fog/off nozzle.

The use of solid streams of water in smaller fires has the disadvantage of scattering burning materials when the stream hits the fire base. The physical damage done by even a medium pressure solid-stream solid hose stream can be considerable. Much time and effort may be required to put out all of any such widely scattered pieces of burning material.

Smaller fires can usually be brought under control by fog and its resulting steam. It is the fog, and the steam it develops, that does most of the extinguishing by cooling and smothering. Once an inert steam atmosphere surrounds flames, little fuel oxidation will be possible, and the flames should decrease. The violent motion of the expanding steam may drive water particles in fog form into confined areas. If these areas are hot when the fog reaches them, the water fog absorbs heat energy, turns to steam, and hopefully stops any flames.

Convection and Ventilation

Convection, the rising of hot gases, is important in structure fires. Consider a fire to be in progress near the floor level of a room. Its smoke is a hot gas. Because of its re-radiation of energy, the smoke molecules push apart, making the gas quite light. As a result, the hot smoke is pushed upward by the heavier, lower, cooler air in the room. These hot convection currents of gas will mushroom outward along the ceiling from the point where they strike it. The amount of heat energy (in btu's) in the room may be staggering. It is not long before such a ceiling surface can be brought up to its kindling temperature.

Suppose a fire involves a living room, burns through into the attic, and starts through the roof. A heavy fog application into the lower area of the room can quench the flames in that area by generating steam, which results in cooling. Some of the warmed fog and steam rises up into the hot attic atmosphere, reducing or stopping the flames there, cooling the under-surface of the roof boards, possibly stopping all flames everywhere, and perhaps in only ten to fifteen seconds. Sounds awfully simple, doesn't it? And sometimes it actually does work that way. But usually fire is not that cooperative, and further follow-up steps must be taken.

The steam developed when water is applied to a fire may actually do much of the extinguishing of the flames.

Note that the coolest area in a room on fire is always near the floor. Anyone trying to get out of any room on fire should bend over and run, or drop to the floor and crawl to safety. Parents and school teachers should expound on this thought: **Down and Out!**

Firefighters without air masks cannot exist in a steam atmosphere (which has a minimum temperature of 212° F, and a humidity of 100 percent). High humidity air at 120° is hazardous, although a dry-air temperature of 200° may be tolerated for short periods of time. Since a single full breath of superheated steam may cause death, it is very important that firefighters wear self-contained breathing equipment.

Up-and-out *ventilation* of a fire scene should be provided as soon as possible to allow hot steam and smoke to rise and clear the room atmosphere and to provide better visibility for the firefighters. In some cases the open doorway of a room on fire may provide sufficient venting. The firefighters in such a case must operate as close to the floor as possible.

Application of excessive water, particularly fog, may overly cool the atmosphere and cause smoke to drop down, making visibility poor.

As soon as ventilation allows visibility in a room, and the fire's dark smoke begins to turn to white condensate, the fog application may be shut down. The nozzle can be changed to provide a narrow cone of spray, directed at any actual burning fuel areas to cool them. The fire is now fairly well contained and is well on its way to being extinguished.

Because heated air and smoke try to rise, opening a vent hole gives them a chance to take their heat away from the fire.

During the containment phase of fire fighting, the goal is to stop the spread of all major flames in one area at a time. There may still be other isolated areas still highly involved.

Only when all visible flames in all areas have been knocked down, when ventilation clears smoke from the area, and when all visible areas that have been involved in flames have been put out by a direct water attack, can the fire be considered as extinguished. Solid streams may have to be used, possibly in short bursts, to drive water into charred areas, cracks, or to deep-seated hot spots to cool them and prevent rekindling.

Solid streams may be used to break out windows to help ventilate a room. They may be able to rip off shingles from a burning roof. At a distance they develop a spray when a fire is too hot for a near approach. If there are no firefighters inside the building, a solid fire-fighting stream can be driven from outside, up through a window or other opening at an inside ceiling. As the water is deflected, it becomes a spray and may be helpful in cooling a fire in higher stories or in attic areas. However, if water is directed into a building through a window, it may push the fire inward and work against containment. More decisions!

There are pro and con arguments whether a firefighter should stand outside and apply a fog cone or solid stream into a window, or whether it would be better to go inside and fight the fire at its source. It is usually better to fight a fire from inside. However, if the building is old and structurally unsound, it should probably be attacked from outside and from a respectful distance.

A brick wall heated on its inner side will expand on that side. The wall may bend until it collapses outward, usually to a distance one- third to one-half its height. When wood heats, it may expand slightly, but a heavily charred wooden surface may contract, which may pull the top of a burned wooden wall inward, toward the fire. But many walls refuse to go by the books. If they are going to fall, they are dangerous!

One of the most effective aids employed by firefighters in producing an eventual extinguishing of active flames is the developing of a vertical ventilation opening high up in a building on fire. This allows natural convective updrafts to help clear out hot smoke and gases. Ventilation does not in itself reduce burning of fuels. In fact, it normally increases the intensity of flame propagation in a structure until fog is applied. What it can do is to produce a natural path for the hot smoke and gases to follow—upward and out. Fresh cool air can then flow into the involved room at the floor level, improving visibility for the firefighters and providing them with a cooler, safer working atmosphere.

Ventilation is sometimes necessary to effect a rescue in a building even before the fire is attacked. A ventilation hole chopped in a roof should be of ample size, at least three by four feet, and directly over the fire area. The rafters should not be cut, or the structural strength of the roof will be lessened considerably.

When setting up a ladder, as for roof ventilation, it should be leaned against the gutter or roof edge some distance upwind of the proposed hole. It should be *footed* (held at the bottom by another firefighter) while it is both pushed up and climbed. It should extend at least three rungs, about three feet, above the roof edge. This enables it to be seen easily and provide a hand-hold when stepping onto the roof from it, or back onto it to descend. At night, a spotlight from the ground should be kept on the ladder top. If the roof slants

steeply, a roof ladder (a ten-foot aluminum ladder with large steel hooks at the top) should be hooked over a peaked-roof ridge between the ascension ladder and the hole. The roof must be tested for structural strength. If it feels springy or spongy when a foot is stamped on it, venting may not be safe. When vents are developed, it is rarely advisable for water to be directed down into a roof vent hole. Fires are best fought from the bottom up.

Horizontal ventilation may be produced by opening up the top section of a window on the lee (downwind) side of a burning structure and opening a door or window on the windward side. Wind, plus the natural upward movement of hot gases, will provide a draft that can carry smoke and gases out through the window opening. To open such a window, it may be necessary to break it from the outside with the top of a ladder, pike pole, axe, or a high-pressure solid hose stream. From inside, even a spray cone directed at a hot window glass will normally break it out. Horizontal ventilation should probably not be used until charged lines are in place. A firefighter and hose may be assigned to guard the areas near a horizontal ventilation hole because the exiting hot gases may ignite wooden areas near the vent hole.

Generally, it is not desirable to open a ventilation corridor through a burning room into and out of an uninvolved room, since this would involve the second room in the fire. However, there may be times when this might be the only choice to take pressure off of some adjacent area to effect a rescue, or if vertical ventilation is not possible. Reliance must be placed on the ability of an immediate fog attack to effectively spread adequate fog and steam into both rooms before damage in the second room becomes excessive. This and many other decisions at a fire scene are only for experienced firefighters to make.

Basement Fires

Basement fires are somewhat different from first or higher floor fires. Water damage to a basement area usually will be less than to living spaces. One of the difficulties is developing ventilation. If there are basement windows, they may be opened. A vent hole may have to be chopped through a basement ceiling to provide an outlet near an external doorway to allow smoke and flames to vent. A rotating cellar nozzle can be run down through any hole chopped in the basement ceiling from the floor above to spray water in all directions around the basement. Possibly the best attack on a basement fire is to use a high-expansion foam to completely flood the whole area with non-flammable, cooling foam.

In a house on fire, if the door knob of a closed-off room is hot, opening the door may cause a "smoke explosion."

Three Possible Fire Phases

If a fire starts in a *completely closed room*, it may go through three phases. At first, with the usual 21 percent oxygen in the air, the fire burns normally. This is called a *first phase fire*. As the oxygen is depleted down to 15 percent, the fire is said to be in its *second phase*, developing heat, some flame, and tarry smoke. Below 15 percent oxygen the fire cannot support flames and is said to be in its *third phase*. It is producing almost no further oxidized chemical heat energy, smoke, or tars. If a fire continues to be supported by only the small amount of oxygen that may seep in under doors or around window frames, the fire may continue to build up heat and to generate carbon monoxide, smoke, and tars. Should a door or a window be opened suddenly into a third phase condition room, there will be the sudden explosive burst of flame, called a *backdraft*, or *smoke explosion*. Windows, doors, walls, and firefighters may be blown outward violently. You, or any firefighter, should ***always*** feel for an abnormally hot doorknob before opening a door to any closed room in a building known to be on fire. If the knob is hot, don't open the door!

In some cases, a third phase fire will smother itself, and eventually the room may cool, and all heat will be dissipated. Everything inside the room will be coated with a layer of

brown tars. Wooden or cloth articles will be charred, and all plastic furnishings will be melted to some extent.

If you should ever come across a closed building that is either on fire or suspected of being on fire, but is showing only wisps of smoke breathing in and out around closed windows and doors, with windows stained brown on the inside, with no flames showing, look out! The situation is ripe for a backdraft explosion. Let the room alone. Backdraft situations require experienced firefighters and are not for the uninitiated to try to handle. In time, the room will probably cool and may then be entered safely.

Figure 13-3. Electric smoke ejector. Brackets are used to wedge unit between door or window frames.

When a fire of any size is finally extinguished, the building will probably need *salvage ventilation*. A unit used for this is a *smoke ejector* (Fig. 13-3). This is a large portable fan pointed downwind into open windows or doorways to blow smoke out of a building. These are most effective if they blow into an open doorway when windows and doors on the other side of the building are opened. It will be found that a 45° fog cone directed out through an open window or door from a couple of feet inside, will also exhaust a surprising amount of smoke from a building. Smoke ejectors can be effectively used during the fire fighting phase if pointed in a downwind direction behind the entering firefighters. Such a forced draft can push smoke and fire out of broken-open downwind windows or open doors, allowing the firefighters to be operating in a cooler area, and can provide better visibility of the fire ahead of them.

Overhaul and Salvage

Overhauling and *salvaging* are important final operations at a fire ground. *Overhauling* means to seek out all hidden hot spots in upholstery, mattresses, behind walls, in window or door frames under sills, in attics, under stairs, under cornices, above hung ceilings, around fireplaces and chimneys, and in basements, to make sure everything is dead out. If there is evidence of arson, such as fire trails or unusual burn patterns, these must be preserved and brought to the attention of a fire ground officer. Any evidence of theft should also be reported.

It is estimated that more than 90 percent of injuries at fires occur during overhaul operations. These are due to the collapse of walls, ceilings, and floors; pulling door and window trim off with pike poles while looking for deep seated areas of fire; by broken glass; by nails in pieces of wood; swinging of edged tools by other firefighters, etc. If the fire occurs at night and visibility is limited, it may be better to postpone the final stages of overhaul until daylight, leaving a watchman to patrol the site. Fires at night usually require many strong, portable flood lights, usually powered by portable auxiliary generators.

All during a fire, it is very important to keep an eye on any nearby exposures (buildings or combustibles of any kind). They may require watering down to keep their surfaces cool. Since infrared radiations will pass unattenuated (without loss) through glass, it may be necessary to enter adjacent exposures and move all combustibles out of the path of such radiant heat waves which might enter through an open or closed window pane. A watch must also be kept downwind for spot fires from firebrands that might land on nearby roofs, in fields, etc.

Salvage operations occur mostly after extinguishment. However, since they minimize losses for the owners of the building, they might

actually be considered to start when a door is sprung open rather than being chopped open, or a less violent fog attack is used rather than a solid stream attack. How the ventilation is carried out, whether it be horizontal or vertical, may determine how much damage is done to a building. Waterproof canvas salvage covers can sometimes be stretched over furnishings and equipment to protect them during a fire, or during overhaul.

Excess water in the building may have to be pumped out. Portable pumps can be used to drain water from cellars, basements, and other areas. Boarding up broken windows and covering roof holes all help prevent further damage from the elements and protect against theft. While salvage operations may not always be a requirement, they can be important factors in improving relations between a fire department and its citizenry.

The absent owners of fire-damaged structures, the building inspection department, and the public utilities are notified of fires by fire department officers. Some of the items requiring immediate attention might be faulty electrical circuits or hanging wires, as well as ruptured or possibly weakened gas or water systems or mains.

Fires in Large Volume Areas

Many stores, supermarkets, and barns are single-story buildings, but will present a different aspect of fire fighting due to their large volume. A small, localized fire in its incipient stage in such structures may be easily contained with a few passes of fog over and into it. When such large areas become heavily involved, great quantities of water will be required because of the large volume of hot gases that can be developed.

The formula to determine the amount of water required to contain a fairly involved fire was given as:

$$\text{gpm} = \text{cu.ft.} \div 100$$

From this, a room of 12 ft. x 14 ft. x 8 ft., or about 1,350 cu. ft., should require a stream of only 13½ gpm to fight a fire in it. A big barn of 50 ft. x 100 ft. x 25 ft., if heavily involved, would require streams totaling 125,000÷100, or 1,250 gpm. This might be possible to fight if city hydrants are available, but may be an impossible situation for most rural fire departments.

If a barn is not too highly involved, it may be possible to walk downwind through the barn with a fog cone, using hay hooks to pull burning bales outdoors where they can be torn apart and drenched. Straight streams may be required to get to burning ceiling areas of barns.

Long steel trusses and beams used to support the roof of wide buildings can be a source of danger. When steel is heated from room temperature to 1,000°F, it will expand about a tenth of an inch for each linear foot of length. Thus, a fifty-foot truss may expand about five inches during a fire, pushing the opposite walls five inches apart before the fire really gets hot. Furthermore, steel beams start to weaken at about 1,100° F, and are likely to collapse when fire heats them to around 1,400° F. On the other hand, wooden trusses expand very little when heated, may remain stronger, and last longer than steel trusses during a fire. They are usually very thick and burn through quite slowly. A straight stream or heavy spray can be used to extinguish any flames which develop on them.

To review, single-story buildings present only five general areas for fire fighting: the basement, first story, attic, rooftop, and some possible exterior additions. The basement may require a large fog attack, possibly a rotating basement nozzle fed down through a hole made in the first floor, or a high expansion foam attack through such a hole, or a doorway. The first floor usually requires a fog, spray, or straight-stream attack, and some form of venti-

lation. If there is no access hole to an attic, the ceiling may have to be pulled down with a pike pole to get into the attic. A fog attack is often all that is required, with no ventilation developed out through the roof unless necessary. A roof surface fire requires a straight stream attack if too far away to reach with a narrow spray cone.

In multi-story buildings, there are all of the considerations that must be given to single-story buildings, plus the difficulties caused by added floors above. In the next chapter, let's see how additional stories may change possible fire-fighting strategies.

14 Multi-Story Structure Fires

Fires in a Two-Story Home

Essentially, all that was discussed about fighting fires in a single-story home or other structure in the previous chapter holds true for multi-story structure fires, such as reporting the fires, the approach, the size-up, rescue activities, laying out hose lines, and how to treat different degrees of fire involvement.

Suppose your home has two stories. If the fire involves only one of the stories, it may be possible to treat it as a single-story fire. But, if the fire is on the ground floor near a stairway leading up to the second floor, hot gases and flames will flow upward and in a short time may involve both floors. The required attack theories will then have to be changed.

In many cases, if a fire has not gained too much headway on the ground floor, an adequate fog attack may develop such a volume of rising steam that it will flow up the stairway and hold down, or, if it has not progressed too far, actually extinguish the beginnings of fire on the second floor.

By developing a vent path by opening a second floor window on the downwind side of a building, and leaving an upwind ground floor door open, a simple fog attack may be all that is necessary to contain a small fire. This can also cause trouble if the fire is a little too intense. The higher a venting opening is in a building, the more effective it may be.

For fires which are not too involved, a starting attack may only require a 2½-inch hose from the pumper, terminated with a double-gated dividing-type siamese wye, feeding two 1½-inch nozzled hand lines going into the house. For larger fires, 2½-inch nozzled lines may be used to apply more water to the areas on fire, but will require more firefighters to handle them.

When a building is seen to be highly involved at arrival, the first thought, of course, is to rescue anyone still inside. Depending on the severity of the involvement, the first floor may be checked by entry through doors or windows. Second and higher floors may have to be entered by ladders if inside stairways are too involved. If a fire escape is attached to a building, a ladder may be leaned against it to provide access to it and the upper floors.

Firefighters must be familiar with seven different types of ladders in their work.

Ladders

The subject of ladders is rather interesting even when not involved in fire fighting. Let's consider some basic information about them. First, most fire equipment ladders are either made of wood, or to reduce their weight, of aluminum. They are much longer and heavier than homeowner ladders. There are seven main types of fire-fighting ladders:

1. Extension ladders, which range from 14 to 60 feet in total length. The longer are carried on ladder trucks.

2. Wall ladders, with no extensions, that range in length from 10 to 35 feet. The longer ones are carried on ladder trucks.

3. Roof ladders, with metal hooks at the top end to hook over the ridge of a roof, range from 10 to 20 feet in length.

4. Attic ladders, that extend from 7½-feet closed to a maximum of about 14 feet. These are used mostly to get to an attic through the small opening often found in the ceiling of a hall or closet.

5. Folding ladders, with rails that fold together to make the ladder slim and easily managed

Suppose the ladder were to be placed against a gutter that extended 18 inches out past the wall, and is 30 feet high—what should the butt distance be from the wall? Figure it out!

CAUTION!

Before raising ladders, check the area above for electric wires or tree branches that the ladder might touch when it is being raised.

in tight spaces. They range in lengths from 8 to 20 feet.

6. *Pompier*, or scaling ladders, have a single 10- to 18-foot vertical spar with rungs through it. The large hook at the top is large enough to hook over a window sill or roof parapet.

7. Aerial ladders, mounted on large, heavy trucks, are rotated and raised by some mechanical means.

Suppose it is required to place a ladder up to a windowsill. Ladders should extend a couple of feet past any window sill through which access is to be gained. For maximum strength, stability, and ease of use, ladders should be set at an angle of about 70° from the ground. A simple formula to determine how far out from the building the ladder butt (base) should be placed is:

$$D = (H \div 5) + 2$$

where D is the distance from the butt to the wall in feet, and H is the height in feet to the point where the ladder touches a window sill or wall. (A little less than 1/3 the height is often a good enough approximation.)

As an example: A windowsill is eighteen feet above ground. The ladder butt should be (18÷5)+2, or 3.6+2, or about 5.6 feet out from the wall. (One-third of eighteen feet is six feet.) If the ladder is to be placed against a roof gutter, add the computed butt-to-wall distance to the distance from the building wall to the gutter point where the top of the ladder will rest.

IMPORTANT! Whenever anyone is using a ladder, there should always be someone at the butt end holding it, unless the butt can be lashed securely to some solid object. Also, the end of a rope that hoists the upper, or fly section of an extension ladder, should be tied tightly around two adjacent lower rungs to assure that the ladder's holding dogs stay locked in position.

To raise long ladders, one person holds the butt to the ground at the desired distance from the building wall. The other person starts at the far end and walks the ladder up to a vertical position. It is then leaned over against the building where it is to rest. With rope-extending type ladders, the ladder is walked up unextended until vertical. It is then extended to the required length before being leaned over to the desired position. At that time the rungs are tied off.

CAUTION! Before raising any ladders be sure to check the area above for electric wires or tree branches that might touch the ladder when it is being raised. With large ladders, four persons may be required to handle them.

Aerial ladders are long extension types, made of wood or metal. They are permanently mounted on a ladder truck. These ladders can be raised, extended, and rotated. Originally this was done by hand, but now it is usually done by a hydraulic system. Some can be extended to sixty-five feet and others to over a hundred feet. These ladders probably operate best at angles of 45° to 60°. When a nozzle is mounted on the fly section, the rig may be called a *ladder pipe* or *water tower.*

Aerial ladders are heavy, and if carrying a load, may tip over any truck on which they are mounted. Their trucks must be quite heavy and as wide as possible. They are outfitted with extendable outrigger jacks on both sides of the truck that provide firm contact with the ground, stabilizing ladder, and truck.

When a platform capable of holding at least two men is attached to the top of an articulating (hinged) boom or pipe and fitted with steps, it forms an *elevating platform*. These are usually big-city types of fire department equipment and are useful for placing water streams into higher story windows, shooting a stream over a parapet onto a roof, rescuing victims in upper stories, providing firefighters direct access to roofs and windows as high as the tenth floor, and breaking out upper story windows for ventilation.

All aerial devices may have problems with icing during cold weather because of spray freezing on them.

The answer to the question above is 9.5 feet.

Fires on Upper Floors

When a fire starts on the second floor of a two-story home or other building, it may often be treated pretty much as a single-story fire. But usually if a building has the lower two or more floors involved in fire, it must be treated as a large fire. Heavy-duty equipment, and a lot of it, will be required. This is the job of trained firefighters, and is really beyond the scope of our book on basic principles and ideas. However, it is a good to have some general knowledge of what is involved to fight such a fire.

Forcible entry means just that—breaking into a building when it is tightly closed. This is often required at fires. Doors or windows of a building on fire must be opened. Be sure to try to turn the knob of a door before breaking it down. Try to push up or pull outward to open a window before prying it open or breaking it in. Some of the forcible entry tools that firefighters use are: axes, bars of various types, battering rams, bolt cutters, hooks, jacks of various kinds, picks, hand saws, and chain or power saws for cutting through walls, roofs, and floors.

Power tools may require electric power from dc generators or ac alternators. Air-operated tools require a compressed air source. While some engines may carry small gasoline motor-driven alternators, large fire departments may have a special vehicle outfitted with gasoline or diesel-driven alternators. Such vehicles may also have air compressors to supply air-operated tools and to replenish firefighter air tanks when they become depleted.

A building having more than a few floors sometimes has a water tank on its roof. Such a tank may be kept filled by being directly connected to the local utility water supply system. One of its two outlet lines may feed all floors with fresh water. A second line from it may feed a 2½-inch or larger *wet standpipe* fire fighting water system. There will be one or more 1½-inch outlets on each floor, on the roof, and in the basement. Single-jacketed, fifty-foot fire hoses will usually be permanently coupled to these outlets in glass-fronted wall cabinets in hallways.

Modern buildings over four stories in height are required to have a *dry, wet* (kept filled with water), or *combination wet-or-dry standpipe system.* At ground level, either inside or against the building wall, will be a wet or dry

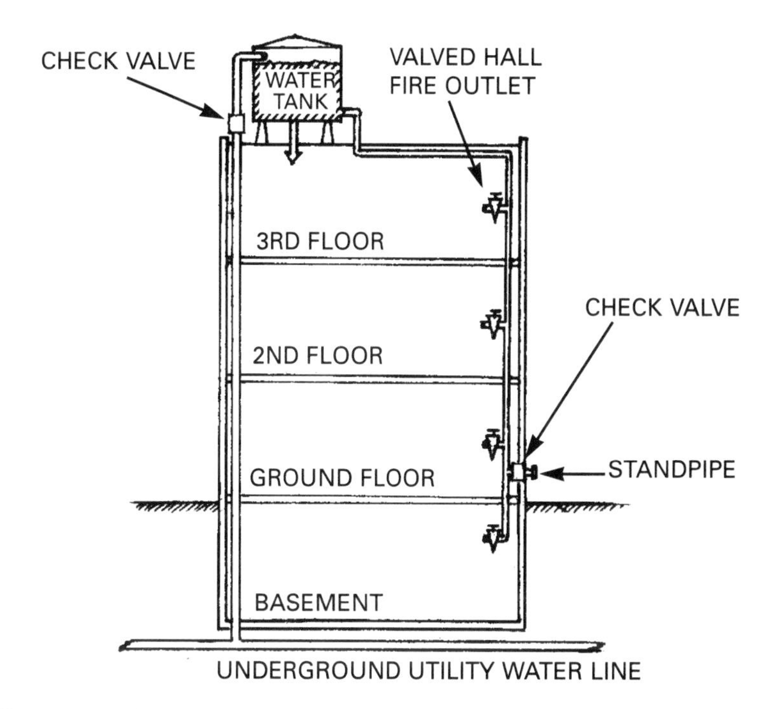

Figure 14-1. Basic standpipe emergency water system for a tall building.

standpipe. A dry standpipe will have no water in it when not in use. Wet standpipes are often found in stairwells and have check valves. A standpipe system may consist of one or more 2½-inch water inlets, mounted a few feet above ground level, or a uniting siamese wye (two-input, one-outlet) fitting into which one or two fire engines can pump water into the building fire fighting water system through a *check valve* (Fig. 14-1). A check valve in a standpipe system prevents water from coming out of the building at the ground level fitting if the pumper shuts down. It will open as soon as the pressure of the water being pumped in by a fire engine is greater than the static water pressure in the building system at that point.

High-rise buildings will usually have an electrical water pump at the ground level of the standpipe system to bring the local utility system's relatively low pressure up to the high

pressure needed for fire fighting. About every ten floors there may be an additional pump to maintain the water pressure for higher floors.

In older buildings many nozzles on the single-jacketed emergency fire hoses installed in building hallways may be only straight stream types. The condition of the hoses may be questionable. Firefighters will usually bring their own heavier-duty 1½-inch single- or double-jacketed hoses with their own on-off/straight-stream/fog nozzles. Newer emergency fire hoses may now be fitted with combination nozzles.

In a burning building never leave any doors open or locked that lead to closed-off hoist shafts, stairwells, or elevator shafts. *Never use elevators in a building that is on fire!*

Today, sprinkler systems are mandatory in buildings frequented by the public. A popular type may have sprinkler heads every ten feet or so across ceilings. Each water outlet may be fed from a ½-inch pipe fitting fed from a ¾- to 1-inch pipe connected to the water system in the building.

The disk, or seat, of a *sprinkler head* is held in place by a metallic strut or a quartz ball between the seat and a yoke that keeps the water turned off. Heated by fire, the lightly soldered strut collapses, or the quartz ball explodes, removing the pressure on the seat. The valve opens and allows water to spray out of the sprinkler head. As soon as the fire is under control, the sprinkler system's main control valve should be shut down. If this valve's location is unknown, the firefighters must insert a wooden wedge into the ruptured head, or possibly use special spring-loaded sprinkler tongs made for this duty. Once the main valve is shut down, a firefighter is usually stationed nearby in case there is a rekindling of the fire. Sprinkler heads should be put back into operational condition and tested before the fire department leaves.

In buildings with many floors, any elevator, air, or hoist shaft, as well as stairways or stairwells, can conduct lower level fire gasses and heat to upper levels. Elevators are very dangerous in fires. For one thing, they may stop working due to the fire or loss of power, sometimes trapping those inside between floors. *Never use elevators in a building that is on fire!*

Stairwell-to-hallway doors in buildings must never be propped open! (They are always spring-closing doors.) During a fire, smoke and gases from lower levels could make the stairwell unusable as a fire escape for anyone at higher levels. A special kind of stairwell fire escape is a *smoke tower.* It is an enclosed stairway with two self-closing doors and a small vestibule between them which prevents any smoke from the burning building from getting into this escape stairway.

In older urban buildings, particularly where tenements or multiple-dwelling units were constructed side by side as one long building, two dangerous fire conditions often existed. Such buildings may not have any horizontal two-by-four fire-break boards added between vertical wall studs. Fire can travel unimpeded from lower to upper floors, or to the attic through such cavities. This is known as *balloon construction.* Feeling walls with a bare hand may tell you if there is fire traveling up inside a wall. If a wall is felt to be hot, the attic should be checked immediately. Also, some long, older buildings may have a common *cockloft,* which is a three- or four-foot attic space between the top floor ceiling and the bottom surface of the roof. In these older buildings, there were no firebreaks built between living units in the cockloft area. Fire could travel rapidly from one end of an attic to the other through a cockloft.

Firefighters may open up a four- to six-foot diameter vent hole in a roof over any top-floor fire to try to prevent the fire from traveling via a cockloft to adjacent units. A ceiling near the opened area might be pulled down with a pike pole to provide firefighter access to the cockloft area from below, or to provide vertical ventilation.

In modern buildings, the difficulties above do not exist because new building regulations require fire breaks between attic sections of multiple dwellings. Firebreak boards are required on all vertical walls, and fireproof insulating materials are now being used between wall surfaces,

between floors, and on top of ceilings, all of which tend to slow down fires.

Large fires in totally involved buildings are sometimes best fought from the downwind side from outside with *master streams* (400 gpm+) of water even though heat, smoke, and flames are at a maximum in this direction. The next best position is from one or both flanks. Surprisingly a poor position is with the wind at your back if you are outside, although this may be the most comfortable for firefighters. Moving downwind is usually the best direction when inside a building.

Fire departments today not only answer fire calls, but are expected to respond to almost any emergency. Let's investigate some of the modern firefighter's non-fire-related activities in our next chapter.

15 Emergencies Other Than Fire

Possible Firefighter Duties

Fire departments are expecting to be called out for all of the emergency situations that are discussed in this chapter, and many others. But you may be able to perform some proper procedures before they can arrive to take over from you.

Originally, fire departments answered only fire calls. Perhaps once in a while, they would be called out to rescue cats that had climbed up tall trees. They might respond to other situations requiring long ladders or streams of water to help the citizens in their area. They could also use their fire engine pumps to drain a basement or two during the rainy season. As a result, a firefighter's life tended to be somewhat spotty. There were short periods of fire fighting interspersed with little to do. They could always polish brass on the rolling equipment, dry hoses, sweep out the firehouse, service fire engines, ladder trucks, and water tenders, and pet or feed their spotted Dalmatian dogs. Much time might go into discussions about what to have for the next meals and who would be the next cook.

It is certainly not so today! Now, fire departments may be overwhelmed with activities. Besides their original fire calls and work, they may also roll on traffic accidents involving gasoline spills and the clean-up that is required. They also respond to many emergency medical calls in their area. Some or all of the firefighters are trained as emergency medical technicians (EMTs). Firefighters will be called if a citizen in their area has a heart attack, has trouble breathing, has serious bleeding, or is downed by an electrical shock.

After earthquakes, floods, hurricanes, bombings, and other disasters, fire departments are always called upon to give a variety of help. EMTs in their fire department vehicle may be called out because they often can get to a destination before an ambulance can. And, of course there may still be the cat rescues as well as using a fire department's high-powered fans to blow unwanted gases out of buildings. Today's EMTs are capable of giving top-grade first aid, applying oxygen masks, or giving *cardio-pulmonary resuscitation* (CPR) and rescue breathing.

Cardio-pulmonary Resuscitation (CPR)

If you should happen to be at the scene of an emergency before a firefighter or an ambulance gets there, you might have to be the one to give the CPR, rescue breathing, or other emergency action. Or perhaps some time you may have to give CPR to a family member or friend. Would you know how to do it? There are some very important first aid procedures that should be mentioned and some of the wide variety of emergencies that today's firefighters may come up against.

It is generally agreed that everyone over twelve years of age should have been trained in CPR and rescue breathing.

It is generally agreed that everyone over twelve years of age should have been trained in both CPR and rescue breathing by the Red Cross. But many people have never attended such training classes. Or, if they have, it may have been so long ago that they are no longer capable of functioning properly in many everyday emergencies. It is recommended that refresher courses of CPR be taken every couple of years. If you have never taken a modern first aid course, do it! Call your local Red Cross for information on the availability of such courses in your area today! If there is no local Red Cross nearby, check with your local fire department for their recommendations. Some fire departments give such training. Learn the ABC's of cardio-pulmonary resuscitation, which are:

A = ***Airway*** *of the victim must be cleared.*

B = ***Breathing*** *must be produced.*

C = ***Compress*** *the sternum bone to pump blood through the victim's heart and system.*

Note: There is sometimes the fear that a rescuing person might become infected with a disease by using mouth-to-mouth resuscitation. This is particularly true of hepatitis, HIV (believed to be the virus that causes AIDS), and others. For this reason, fire departments may use plastic one-way breather valves inserted into the victim's mouth, or they may use a device, such as a squeezable air bag. Ambulances use automatic breathing machines.

There may be little to fear from one's own children and family members who are not known to have any infectious diseases. *Mouth-to-mouth rescue breathing* is:

1. *Alternately breathing into a victim's mouth to fill his or her lungs with air from your lungs, while holding his or her head back with nose pinched closed, and then*

2. *Removing your mouth to let the person exhale your air.*

If there is a question in your mind, it might make you feel better if you hold a handkerchief over your own mouth when you breathe into the other persons mouth, pull your mouth and the handkerchief away when the other person is exhaling.

Heimlich Maneuver

The *Heimlich maneuver* may sometimes be used to try to restart breathing or to start a stopped heart.

If a person is choking on food, (indicated by grasping the throat and being unable to speak), have someone call 911 (which means the closest fire department). Say to the struggling person, "Can you talk to me?" If not, it is recommended that you position yourself to the left of the person, with your left arm tight against the victim's stomach. Have him or her bend forward against your arm. Give four hard blows with the heel of your open right hand to the victim's back, between the shoulder blades. (If the victim can speak, then the trouble is not in the airway, and it may be something like a heart attack. In that case, you would not give any blows.) If the four blows do not dislodge the food, and the person is still unable to breathe or speak, take the next step.

Get behind the person, make a tight fist with one hand and place it three inches beneath the victim's sternum or breast bone. Holding the fist-hand wrist with your other hand, push your fist in and up in several quick, hard thrusts (this is the Heimlich maneuver, or procedure). In effect, you are trying to cause the victim to cough very hard to dislodge the food. If this produces no results, see if you can dislodge any food from the throat or air passage with your finger. If this is still not successful, repeat the four blows, and then try the Heimlich maneuver again. Continue until the person starts breathing or becomes unconscious. In the latter case, continue the same operations with the person lying on the floor until the fire department or medical help arrives. ***If a victim does not breathe, he or she will die!*** Continue to do anything you can that might start the breathing again. You must understand that this book can give you only some good ideas. You should take accredited training courses to learn how to carry out properly any of the ideas that are outlined!

Automobile Accidents

Automobile accidents often present serious problems to rescuers. Injuries to drivers or passengers must be taken care of first, of course. An attempt should be made to get passengers out of vehicles if there is any chance of a fire occurring due to spilled gasoline. Beware of the possibility of an explosion if a spark or a flame should reach gasoline fumes from a ruptured gas tank or other spillage onto the ground. Make sure that no one is smoking in such an area or doing anything else which might produce a

spark, such as hitting steel against steel. A portable hand extinguisher may be useful on fires under the hood or inside the car.

The spilling of gasoline onto roadways or into gutters requires immediate action by a fire department. Care must be taken to prevent gasoline from entering gutter sewer drains. Gasoline fumes in a sewer can form a highly explosive mixture and result in great damage if ignited. While the range of the percentage of gasoline vapor to air-oxygen which will support an explosion is relatively small (from 5 to 10 percent gas-to-air), it is remarkable how many times such an explosive combination occurs.

A preferred method of preventing the possible contamination of ground water with spilled gasoline or oils on roadways is the spreading of *rice hull ash* over the spill. This absorbs the gasoline or oil. The residue must be swept up, bagged, and removed to a safe dump site. Although diesel and other oils are less flammable than gasoline, on a street they can produce a dangerously slick surface as well as contaminate the environment.

When people are trapped inside a vehicle, it may be necessary to use the *Jaws of Life* rescue equipment. With this group of pneumatic or hydraulic cutting, expanding, and pulling tools, the top, doors, windshield, windows, or sides of trucks or automobiles can be cut, pulled, or pried apart to allow rescue of persons inside of vehicles.

Because of the great weight and strength of fire engines or water tenders, in an emergency they may be called upon to drag cars out of streams or other places to allow rescuers to get to people inside.

Electrical Emergencies

Electricity, even at the voltages around us daily, can be deadly. A person can be electrocuted with the normal 120-volt ac in any wall outlet socket if the person makes a solid enough contact between the two lines or gets between ground and the electrically *hot* wire. Luckily, in most cases, people make only a short-duration, brushing contact across such lines. The result is usually a sharp muscle reaction, possibly resulting in the person being thrown backward or perhaps off a ladder. In many cases, people are hurt more by the physical reaction contact against some hard object, or by a resulting fall, than by the electrical shock itself.

What should be done if someone is rendered unconscious by an electric shock? Have someone call 911. This can be a dangerous situation for any rescuer. If the victim is still on or touching electric wires, try to disconnect the electricity from those wires by pulling a wall plug or by throwing the main switch at a distribution box to the OFF position. Although it is extremely dangerous, it is sometimes necessary to try to pull a person away from a live wire by his or her clothing. This must be done only if the clothing is dry. Sometimes victims may be pushed or pulled off a hot wire with a dry wood or plastic pole. Be sure that releasing a person from a downed line does not allow the hot wire to spring over against you! Try to anchor the hot line by tossing something heavy across it first. Once the victim is clear of the hot line, check for breathing and for any pulse. If neither, give rescue breathing and/or CPR.

Be very careful about the possibility of getting electrical shocks if you are near water, metal water pipes, or wet surfaces, such as in bathrooms, kitchens, laundry rooms, or near old outside switch boxes. At emergency scenes, electrical wires hanging down or lying on the ground, may be very dangerous. Keep everyone away from them even if you think they are probably not hot.

Get passengers involved in accidents out of vehicles if there is any chance of a fire occurring due to spilled gasoline.

Be careful of electrical shocks if you are near water, metal water pipes, or wet surfaces in bathrooms, kitchens, laundry rooms, or near old outside switch boxes.

Drowning

When faced with a drowning, have someone call 911. If rescue from the water is required, it is imperative that you not be pulled under the water by the frantic grasping by the drowning victim. When the victim is out on dry ground and coughing, stand by and see what happens. Often the victim can recover unat-

tended. If the victim is not breathing and there is no pulse, start immediate CPR and rescue breathing. A victim will die or be brain-damaged if the heart does not pump blood to the brain for only four to five minutes. Do not waste any time starting CPR!

As soon as possible, try to get the drowning victim wrapped in blankets to maintain body temperature. If a victim is rescued from near-freezing water, he or she may have a very low body temperature. It is interesting that a successful revival of heart operation has been known to occur as much as a half-hour after rescue—so don't give up too soon! When a person is not breathing and no pulse can be felt, the Heimlich maneuver might be tried. It compresses the lungs to squeeze water out of them and also compresses and releases the heart to try to start it pumping if it has stopped.

In the case of a broken bone, call 911, but do not move the victim until help arrives. Apply pressure with a cloth bandage over any area of bleeding.

Fractures

In the case of a broken leg, arm, or other bone, have someone call 911, but do not move the victim until help arrives unless there is immediate danger of further injury. Apply direct, reasonably hard pressure with a cloth bandage over any area of bleeding. Keep the victim calm and as warm as possible. Do not push broken bones back into place if they are protruding from the skin. Splint unstable fractures with a pillow or boards, as recommended by the Red Cross course you should have taken, to prevent painful and damaging movement.

Bleeding from an ear may mean that the skull is fractured. Have someone call 911. Most scalp wounds bleed profusely. Press only very lightly on head wounds with any type of clean cloth bandage to prevent possible bone fragments from being pressed into the brain area. Always suspect a neck injury with a serious head wound. Immobilize the head and, if possible, do not move the victim until help arrives.

Drug Overdose

Drug overdose is a form of poisoning. Call 911 at once. Check for a pulse and breathing. If no pulse is present give CPR. If no breathing is detectable, give rescue breathing. Remember that deadly infectious diseases are always a possibility with drug addicts, particularly those who use needles to inject drugs. When victims are revived, keep them warm by covering with a blanket until firefighter or ambulance help arrives.

Burns

If someone is severely burned, call 911. It is usually recommended that burns be cooled with water. Carefully remove any clothing from a badly burned skin area if it is possible to do it without pulling off the skin itself. Clean sheets or towels may be used to cover the burned areas during transportation to a hospital. For chemical burns, wash with clean running water for twenty minutes or until medical help arrives. If chemicals have splashed into the eyes, flush them with clean water for twenty minutes or until help comes.

Smoke Inhalation

Anyone suffering from severe smoke inhalation may require either rescue breathing or CPR (or both) in severe cases. Have someone call 911. The victim should be taken to a medical facility for observation. It may take several hours for dangerous symptoms to appear.

Seizures

A seizure is quite an alarming thing to witness. Have someone call 911. The victim thrashes about and in so doing may injure himself or herself. There is not much that you can do except to keep objects out of the way that might cause injury if struck by the thrashing victim. It is usually recommended that you allow the seizure to run its course and then make the victim as comfortable as possible until medical help arrives.

spark, such as hitting steel against steel. A portable hand extinguisher may be useful on fires under the hood or inside the car.

The spilling of gasoline onto roadways or into gutters requires immediate action by a fire department. Care must be taken to prevent gasoline from entering gutter sewer drains. Gasoline fumes in a sewer can form a highly explosive mixture and result in great damage if ignited. While the range of the percentage of gasoline vapor to air-oxygen which will support an explosion is relatively small (from 5 to 10 percent gas-to-air), it is remarkable how many times such an explosive combination occurs.

A preferred method of preventing the possible contamination of ground water with spilled gasoline or oils on roadways is the spreading of *rice hull ash* over the spill. This absorbs the gasoline or oil. The residue must be swept up, bagged, and removed to a safe dump site. Although diesel and other oils are less flammable than gasoline, on a street they can produce a dangerously slick surface as well as contaminate the environment.

When people are trapped inside a vehicle, it may be necessary to use the *Jaws of Life* rescue equipment. With this group of pneumatic or hydraulic cutting, expanding, and pulling tools, the top, doors, windshield, windows, or sides of trucks or automobiles can be cut, pulled, or pried apart to allow rescue of persons inside of vehicles.

Because of the great weight and strength of fire engines or water tenders, in an emergency they may be called upon to drag cars out of streams or other places to allow rescuers to get to people inside.

Electrical Emergencies

Electricity, even at the voltages around us daily, can be deadly. A person can be electrocuted with the normal 120-volt ac in any wall outlet socket if the person makes a solid enough contact between the two lines or gets between ground and the electrically *hot* wire. Luckily, in most cases, people make only a short-duration, brushing contact across such lines. The result is usually a sharp muscle reaction, possibly resulting in the person being thrown backward or perhaps off a ladder. In many cases, people are hurt more by the physical reaction contact against some hard object, or by a resulting fall, than by the electrical shock itself.

Get passengers involved in accidents out of vehicles if there is any chance of a fire occurring due to spilled gasoline.

What should be done if someone is rendered unconscious by an electric shock? Have someone call 911. This can be a dangerous situation for any rescuer. If the victim is still on or touching electric wires, try to disconnect the electricity from those wires by pulling a wall plug or by throwing the main switch at a distribution box to the OFF position. Although it is extremely dangerous, it is sometimes necessary to try to pull a person away from a live wire by his or her clothing. This must be done only if the clothing is dry. Sometimes victims may be pushed or pulled off a hot wire with a dry wood or plastic pole. Be sure that releasing a person from a downed line does not allow the hot wire to spring over against you! Try to anchor the hot line by tossing something heavy across it first. Once the victim is clear of the hot line, check for breathing and for any pulse. If neither, give rescue breathing and/or CPR.

Be very careful about the possibility of getting electrical shocks if you are near water, metal water pipes, or wet surfaces, such as in bathrooms, kitchens, laundry rooms, or near old outside switch boxes. At emergency scenes, electrical wires hanging down or lying on the ground, may be very dangerous. Keep everyone away from them even if you think they are probably not hot.

Be careful of electrical shocks if you are near water, metal water pipes, or wet surfaces in bathrooms, kitchens, laundry rooms, or near old outside switch boxes.

Drowning

When faced with a drowning, have someone call 911. If rescue from the water is required, it is imperative that you not be pulled under the water by the frantic grasping by the drowning victim. When the victim is out on dry ground and coughing, stand by and see what happens. Often the victim can recover unat-

tended. If the victim is not breathing and there is no pulse, start immediate CPR and rescue breathing. A victim will die or be brain-damaged if the heart does not pump blood to the brain for only four to five minutes. Do not waste any time starting CPR!

As soon as possible, try to get the drowning victim wrapped in blankets to maintain body temperature. If a victim is rescued from near-freezing water, he or she may have a very low body temperature. It is interesting that a successful revival of heart operation has been known to occur as much as a half-hour after rescue—so don't give up too soon! When a person is not breathing and no pulse can be felt, the Heimlich maneuver might be tried. It compresses the lungs to squeeze water out of them and also compresses and releases the heart to try to start it pumping if it has stopped.

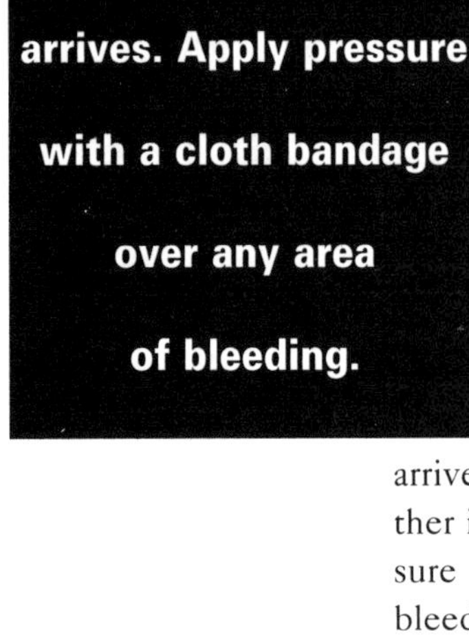

Fractures

In the case of a broken leg, arm, or other bone, have someone call 911, but do not move the victim until help arrives unless there is immediate danger of further injury. Apply direct, reasonably hard pressure with a cloth bandage over any area of bleeding. Keep the victim calm and as warm as possible. Do not push broken bones back into place if they are protruding from the skin. Splint unstable fractures with a pillow or boards, as recommended by the Red Cross course you should have taken, to prevent painful and damaging movement.

Bleeding from an ear may mean that the skull is fractured. Have someone call 911. Most scalp wounds bleed profusely. Press only very lightly on head wounds with any type of clean cloth bandage to prevent possible bone fragments from being pressed into the brain area. Always suspect a neck injury with a serious head wound. Immobilize the head and, if possible, do not move the victim until help arrives.

Drug Overdose

Drug overdose is a form of poisoning. Call 911 at once. Check for a pulse and breathing. If no pulse is present give CPR. If no breathing is detectable, give rescue breathing. Remember that deadly infectious diseases are always a possibility with drug addicts, particularly those who use needles to inject drugs. When victims are revived, keep them warm by covering with a blanket until firefighter or ambulance help arrives.

Burns

If someone is severely burned, call 911. It is usually recommended that burns be cooled with water. Carefully remove any clothing from a badly burned skin area if it is possible to do it without pulling off the skin itself. Clean sheets or towels may be used to cover the burned areas during transportation to a hospital. For chemical burns, wash with clean running water for twenty minutes or until medical help arrives. If chemicals have splashed into the eyes, flush them with clean water for twenty minutes or until help comes.

Smoke Inhalation

Anyone suffering from severe smoke inhalation may require either rescue breathing or CPR (or both) in severe cases. Have someone call 911. The victim should be taken to a medical facility for observation. It may take several hours for dangerous symptoms to appear.

Seizures

A seizure is quite an alarming thing to witness. Have someone call 911. The victim thrashes about and in so doing may injure himself or herself. There is not much that you can do except to keep objects out of the way that might cause injury if struck by the thrashing victim. It is usually recommended that you allow the seizure to run its course and then make the victim as comfortable as possible until medical help arrives.

While on the subject of safety around the home or work place, there are three emergencies for which some basic suggestions might be made, about how best to react to them, or what to do to prepare for them:

Earthquakes

Heavy things that fall or slide around in rooms are major dangers. In earthquake country, all interior water tanks and other heavy items, such as stoves, filing cabinets, desks, etc. should be strapped or anchored to studs in the walls next to them. Positive latching cabinet doors should be used. If you can safely do it, try to get to a clear area outside. But stay away from power lines that may come down, as well as brick chimneys or walls, which could fall. If there is a brick chimney next to an exterior door, do not use that doorway to escape from the building during an earthquake. Heavy items on high shelves are dangerous in earthquake country. Expect everything on shelves to be on the floor after a really good shake. Keep all heavy items on the lowest shelves. Inside a building, one of the safest actions in a quake may be in stand in, and hold onto, a strong open doorway. Ducking under a sturdy table or desk and holding onto something may also give protection from falling and sliding objects. Keep away from glass windows and cabinets. When the shaking stops, check for injured persons first, then turn off the house electricity and gas mains to prevent possible fires, and the water main if there are broken pipes. Much of the damage from earthquakes is due to fire! Try to extinguish fires before they gain headway. Call 911 and hope that the telephone lines are not down. Don't forget, if you have a cellular or other radio telephone, it may operate when your home phone won't. Always prepare for aftershocks! Turn on a portable radio for emergency information and damage reports. Have spare batteries on hand for radios and flashlights. Keep at least ten gallons of drinking water in storage at all times.

Explosions

A large explosion nearby may act somewhat as an earthquake. Dive under a table or desk. Always consider smoke to be poisonous. Try to move to the upwind side of any explosion area. If you are downwind and cannot get away, an airtight interior room might be best to stay in until the gases clear. Check for burning firebrands on or near your property, and try to extinguish them while someone calls 911.

Hurricanes or Tornadoes

Forewarning is usually provided by radio and TV. Buildings may require the protective covering of windows with plyboards. Buildings are usually most rigid and strong if all of the interior doors and windows are completely shut. If you are not going to attempt to escape the area, a basement or other closed-in, low, or underground place should be the safest. If these are not available, find an inside hallway or other small area that has no windows and four walls, with doors that close, to protect you from flying objects that my be driven through outer walls or windows. Close all doors to your safety area. Sitting on the floor and surrounding yourself with blankets and mattresses will help protect you. Turn off the electricity and gas mains to prevent fires and the water main in case pipes become broken. The professionals to contact for all types of emergency information are the Red Cross, police, and fire departments in your area. Call them about their recommendations.

Hazardous Materials

Fire departments today are called out on many hazardous material incidents. Trained firefighters may try to identify, contain, and then clean up the material in simpler spills, or have a special *Hazmat* (hazardous materials) team attempt to make an identification of any suspected hazardous substance. These materials must be disposed of by the Hazmat team, rarely in local dumps. Some fire departments even have chemists who accompany the firefighters on Hazmat calls.

People call the fire departments for all kinds of assistance. They call for smoke investigations, unidentified smells, gas leaks, explosions, automobile accidents, downed power lines, water leaks, floods, tornadoes, earthquakes, industrial accidents, elderly people needing emergency assistance, and so on.

The fire chief of a city or rural area has his or her fingers on the pulse of the area. As a result, fire chiefs usually work closely with the planning departments of their local jurisdictions. Some of the firefighters may be trained to make periodic safety inspections of buildings in the area. Such firefighters must also have a thorough knowledge of local building and health codes.

The information above is really only a brief outline of some of the things that may be required of firefighters, and possibly you, if you happen to be at the scene of an emergency first. Again, for emphasis, you and your family should be properly prepared—you and they should all take the first aid and CPR courses given by the Red Cross. A good outline of first aid procedures, including CPR, may be found in the front of your telephone book. Check this out. Read it carefully and discuss it with your family.

It can be seen that the work of firefighters is not just dumping water on a fire. They must be well-trained in many areas and be capable of responding in a proper manner at any time to a wide variety of life-threatening emergencies to protect the citizens in their area. The debt that we owe such dedicated people working for our well-being cannot be over-emphasized.

16 Home Fire Safety Tips

How to be Prepared

Once again, every home, particularly if it is in the country or in suburbia, should have ready at all times at least one (two, for front and back yards) all-weather, flexible, non-kinking plastic hose, at least 75 feet long, with a ⅝-inch inside diameter, terminated with an adjustable/off garden-hose nozzle permanently screwed onto its male-threaded end. Keep it in a 3-foot diameter roll in an easily accessible place. Leave it permanently attached to a centrally located exterior hose faucet during hot, dry seasons. When it is not connected, it should be rolled up and stored where it will be easily available. While stored, be sure to push a cork loosely into the open female-threaded end of the hose so that bugs cannot crawl inside. They would foul the nozzle when water is fed through the hose at a fire. Close off the nozzle for the same reason. If you live in an area subject to subfreezing temperatures, make sure the hose is stored completely empty during the cold months. While connected, keep the hose in a shaded area if possible, and test for proper water flow and nozzle operation each month.

Using the hose described above, with 50 psi from a home water system, you should be able to throw a reasonably solid 3⁄16- or ¼-inch stream of water horizontally about forty feet, and a 20° spray at least twenty feet. From the ground, the solid stream should reach most of the roof area of small or medium-sized single-story homes. If used while on the roof, or in a second-story building that is about ten feet above the faucet, the pressure (assuming the nozzle is being held five feet above the floor level) will be reduced 0.43 x 15, or by about 7 psi, leaving a working pressure of 43 psi. Even if operated on a third floor, the water pressure will still be about 36 psi, which can deliver a reasonably usable spray for small fires in chairs, couches, etc.

Keeping Your Home Safe

Besides assuring yourself that you have a relatively useful water supply to use against small home fires, there are other things that will provide protection against the setting of, or extension of, fires. Many of these suggestions have been explained elsewhere in this book.

1. If in a rural area, keep vegetation, or anything that might be combustible, cleared away from your home by at least 30 feet.

2. Remove the bottom 6 feet of all branches growing on tall trees to prevent ground fires from extending up into the tree branches.

3. Remove all flammable debris from your basement, attic, unused rooms, garage, and yard.

4. Stack woodpiles on cleared areas at least 15 feet from all structures.

5. Install heating/cooking-gas tanks at least 50 feet *(check your local requirements)* from any normal home, unless its nearest wall is fireproof and has no window or door openings. Keep the tank area cleared of all vegetation for at least 10 feet in all directions.

6. If burn barrels are used, keep them in a large, cleared area and use a ¼- to ½-inch mesh cover on them when in use to prevent firebrands from flying out onto anything combustible.

7. Keep tools that might be useful in case of a fire, such as buckets, shovels, rakes, etc., in a handy place.

8. Every time you leave the house for any length of time, make sure that all exterior windows, and most importantly, all of the interior doors, are tightly closed. Remember, if a fire starts in an airtight room it may put itself out and limit the fire to that one room, when the air-oxygen in the room is used up by the fire. (And do this every night when you go to bed!) This may be the simplest but most important thing you can do to reduce the possibility of the burn-down of your home!

9. Mount smoke alarms on ceilings in hallways outside of bedrooms, kitchens, at tops of stairways, and other places in homes. If anyone smokes in bed, be sure to have a smoke alarm mounted *in that bedroom.* These devices have saved thousands of lives since they were first developed. Replace smoke alarm batteries twice a year (when clocks are changed from standard to daylight savings times is a good plan). *Your fire department, if asked, will usually advise you as to the best places to mount your alarms.*

10. Keep your chimney flue clean and check to see if any attic or outside woodwork near the chimney appears to be overheating. If so, repairs to the chimney are indicated.

11. Keep a ¼- or ½-inch mesh cap over the top of your chimney flue to prevent sparks from falling onto your roof or onto nearby combustible materials.

12. If you ever notice an electrical switch or its plate is warm to the touch, try tightening the internal switch-contacts (often necessary if aluminum wire is used in the house wiring). If the screws are found to be tight, the electrical load on the line may be excessive, or the switch contacts may not be making good connections, which would require a new switch. Before working on any switch or other electrical device, *be sure the electricity is disconnected* by opening the circuit breaker to that line (removing the fuse, in older homes).

13. Never jumper or do anything to defeat a circuit breaker or fuse. Have an electrician check any circuit in which breakers or fuses repeatedly open.

14. Keep drapes or other combustible materials away from your TV set.

15. Make sure you and all of your family know how to turn off the electricity, water, and gas at the outdoor entrance switch boxes and valves.

16. Store all flammable liquids in safe places and in metal containers. *(Glass containers might break, and some plastics might weaken due to material fatigue or chemical reaction with the liquids.)*

17. Make sure that your electric and/or gas heating and cooking equipment is always in good repair to prevent electrical shorting and sparking, or unexpected ignition of leaking gases.

18. Keep loose clothing and other flammable materials away from open flames.

19. Keep a fire screen in front of open fireplaces when in use. Make sure fireplace-insert doors are always closed, and wood-burning heater/stoves are on a large metal or fireproof mat.

20. Teach children not to play with matches and not to get near fireplaces and barbecue equipment when these devices are in use.

21. Teach children to keep flammable articles away from gas or electric heaters or stoves, and from lighted candles or gas lamps.

22. Keep oily or greasy rags in metal containers because they may break into flames due to *spontaneous combustion.* (Oily rags oxidize very slowly, but if held in tight containment, they may accumulate chemical heat energy until the ignition point is reached, at which point they will break into flames.)

23. Make sure your electrical outlets are in proper operating condition and are not warm to the touch. *(Use child-safe sockets on all wall outlets that children can reach.)*

24. Never overload electrical wall outlets.

25. Do not use extension cords if they feel warm when in use.

26. Never run extension cords under carpets where they can be stepped on.

27. Do not use electrical appliances near faucets or other grounded objects unless *ground-fault circuit breaker outlets* are used in kitchen, bathroom, laundry, or unfinished basements.

28. Use only CO_2, or other Class C extinguishers on electric fires, and Class B or C extinguishers on grease or oil fires in the kitchen *(Chapter 5).*

29. Check often that all doors, windows, and screens will operate and open easily to assure easy exit for anyone wanting to leave the home if it is on fire.

30. Designate some meeting place on your property where everyone living in the home will gather in case of a fire, so you will know when all residents are out of the building.

31. When dumping fireplace ashes, use metal cans and allow at least thirty-six hours for the ashes to cool. Often, twelve-hour, or older, ashes will appear to be out, but live coals will be hidden in them. Fires can be started by such coals.

32. Since home fires occur most often during winter months, think most about fire in the home during this time of the year.

33. If your home has safety bars on windows, install safety-release latches on the bars. *(This is very important! Contact your police or fire department for information on this subject.)*

34. Be sure all persons in the home know how to open all of the safety locks on all doors and windows.

35. If you find it necessary to replace your roof, be sure to use only *fireproof* roofing materials.

36. Be careful with automatic coffee makers and kitchen stoves—they have caused many fires because they were not turned off properly.

37. Do not burn garbage or newspapers in your fireplaces. Have your chimney inspected for creosote and possible leaks every few years.

38. If you are unable to get a small fire under control with a fire extinguisher or a hose within a few seconds, have someone call 911. *Get those fire engines rolling toward you as soon as possible!* You can always cancel the call if you are successful in putting out the fire.

Home Fire Drills

• An excellent exercise for homes, particularly those in which there are small children or elderly persons, is to have a home fire drill, perhaps on the first day of each month (so it is easy to remember). Sound a horn, bell, or whistle, and have everyone exit the house and go through the motions of producing some kind of a fire response.

• Have all young children demonstrate their ability to dial 911. (Of course, you must hang up the phone before the last number is dialed to prevent a false alarm.)

• Connect to a faucet, and even turn on, an outdoor hose that could be used to put out a fire.

• If your home has two stories, have at least one rope ladder (one for each room would be better) by which anyone upstairs can get safely to the ground in a fire emergency. Practice anchoring the top end and lowering the other end out a different window or over a different porch railing each month.

• Time your drills. Make a game out of them to see how little time your fire drills take each time. Get kids thinking about the dangers of fire and even how they might be able to extinguish one!

• In emergencies, if clothing catches on fire, wrap a small throw rug, blanket, or coat around the victim. If it happens to you, remember to SDR (Stop, Drop, and Roll). You might wrap a small throw rug around yourself or throw yourself on a bed and roll the bedding or a blanket around you as tight as possible to smother the flames. Teach your whole household how to do this.

• Burns on most parts of the body are usually best treated by pouring cool, clean water over the burned area for several minutes. Get emergency aid as soon as possible if burns are extensive.

Stay Informed

• Fires cause great losses of property and lives as well as much anguish every year. Remember, most fires should not have occurred. Someone either illegally set them on purpose, or someone did something foolish that allowed the fire to start. If we constantly consider the fire implications of what we are doing, it will be unlikely that we will be guilty of starting any fires. Constantly work on that idea. When you are faced with a fire emergency some time in the future, let's hope that something that this book taught you will help you to respond properly to that emergency.

• If you think you would like to try to become a firefighter, contact the fire chief or training officer of your local fire department. Tell them what you have learned about fighting fires.

INDEX

Notes

Notes

Notes

Notes

Notes